Union Public Library
1980 Morr'
Union, N.J.

Decoding Gardening Advice

Decoding Gardening Advice

THE SCIENCE BEHIND THE
100 MOST COMMON RECOMMENDATIONS

Jeff Gillman & Meleah Maynard

Union Public Library
1980 Morris Avenue
Union, N.J. 07083

Timber Press
PORTLAND · LONDON

Copyright © 2012 by Jeff Gillman and Meleah Maynard.
All rights reserved.

Published in 2012 by Timber Press, Inc.

The Haseltine Building
133 S.W. Second Avenue, Suite 450
Portland, Oregon 97204-3527
timberpress.com

2 The Quadrant
135 Salusbury Road
London NW6 6RJ
timberpress.co.uk

Printed in the United States of America

Library of Congress Cataloging-in-Publication Data

Gillman, Jeff, 1969-
  Decoding garden advice : the science behind the 100 most common recommendations /
Jeff Gillman and Meleah Maynard. – 1st ed.
      p. cm.
  Includes bibliographical references and index.
  ISBN 978-1-60469-220-4
  1. Gardening. I. Maynard, Meleah. II. Title. III. Title: Good and bad gardening advice
and how to tell the difference.
  SB450.97.G55 2011
  635–dc22
                      2011013138

A catalog record for this book is also available from the British Library.

For Suzanne,

Catherine, and Clare

For Mike,

who did everything that needed doing

while I wrote

CONTENTS

Acknowledgments 8

Introduction 9

1 SOIL..12

Organic material ∘ Earthworms ∘ Vermicompost ∘ Soil tests ∘ Lime ∘ Sulfur ∘ Fertilizing and the weather ∘ Soil in containers ∘ Urine as fertilizer ∘ Fertilizing in spring ∘ Peat moss ∘ Yearly fertilizing ∘ Tilling ∘ Organic versus synthetic fertilizers ∘ Ammonium and nitrate ∘ Bacteria and beans ∘ Mycorrhizae ∘ Balanced fertilizers ∘ Pine needles ∘ Compost tea ∘ Sand

2 WATER..54

Watering depth and frequency ∘ Planting trees and shrubs ∘ Time of day ∘ Checking soil moisture ∘ Overhead watering ∘ Watering trees ∘ Withholding water ∘ Drainage in containers

3 PEST, DISEASE, AND WEED CONTROL....................68

Insecticidal soaps ∘ Timing of pesticide application ∘ Homemade deer repellents ∘ Organic versus synthetic pest control ∘ Homemade insect sprays ∘ Corn gluten meal ∘ Ladybeetles ∘ Glyphosate ∘ Controlling small mammals ∘ Pesticide shelf life

4 MULCH 86

Organic versus inorganic mulches ∘ Mulch placement ∘ The necessity of mulching ∘ Winter mulch ∘ Landscape fabric ∘ Wood mulch ∘ Nitrogen and wood mulch

5 ANNUALS, PERENNIALS, AND BULBS......................102

Deadheading ∘ Hardiness zones ∘ Hardening off seedlings ∘ Root-bound plants ∘ Planting non-natives ∘ Perennial beds beneath trees ∘ Disease-resistant cultivars ∘ When to plant perennials ∘ Spacing recommendations ∘ Grow lights ∘ Light levels ∘ Sterilizing containers ∘ Dividing and transplanting ∘ Phosphorus

6 TREES AND SHRUBS 134

Dividing shrubs ∘ Pruning ∘ Planting depth ∘ Planting hole size ∘ Soil level ∘ Fertilizing ∘ Staking ∘ Tree wraps ∘ Wrapping evergreens ∘ Balled-and-burlapped trees ∘ Topping trees ∘ Dressing pruning wounds ∘ Pruning at planting ∘ Beating trees ∘ Pruning cuts ∘ Shearing deciduous shrubs

7 VEGETABLES AND FRUIT 162

Pruning fruit trees ∘ Eggshells and tomatoes ∘ Crop rotation ∘ Thinning seedlings ∘ Weeding ∘ Hybrid versus nonhybrid seeds ∘ Treated lumber ∘ Planting in rows ∘ Lunar gardening ∘ Companion planting ∘ Vegetables in containers ∘ Light levels

8 LAWN CARE...188

Dog spots ∘ Yard waste disposal ∘ Mower height ∘ Cutting grass ∘ Seeding ∘ Low-input grasses ∘ Sod versus seed ∘ Borax ∘ Light levels ∘ Lawn chemical safety ∘ Fertilizers and pollution ∘ Organic fertilizers ∘ Grass clippings ∘ Brown lawns ∘ Synthetic fertilizers and insecticides ∘ Overwatering

Conclusion 216

Helpful Conversions 217

Selected Bibliography 218

Index 219

ACKNOWLEDGMENTS

Many thanks to all of the kind and generous master gardeners, farmers, horticulturists, professors, garden writers, entomologists, and researchers who offered opinions, advice, and their own stories of gardening successes and failures. Your humor and wisdom helped shape this book, and we are deeply grateful. Special thanks to master gardener Theresa Rooney, whose encyclopedic gardening knowledge and good cheer are unparalleled. Thanks, as always, to Chad Giblin for helping Jeff out while he was off writing. We also deeply appreciate the thoughtful work of our copyeditor, Lisa DiDonato Brousseau.

INTRODUCTION

As an associate professor of horticulture at the University of Minnesota (Jeff) and a master gardener (Meleah), we field questions all the time about everything from watering and soil preparation to fertilizing, pruning, and chemical use. We didn't realize how many dos, don'ts, and shoulds there are to gardening until people started asking us these questions, half of which we couldn't answer without poring over garden books and research papers ourselves.

One afternoon we were talking about the sorts of things we're asked most often, and we were shocked to realize how much advice gardeners at all levels are bombarded with. Who could possibly make sense of it all? What bothered us most was knowing that the confusion caused by this mountain of sometimes contradictory advice may turn gardening into a joyless, by-the-numbers exercise. Some would-be gardeners get scared, thinking they will never figure it all out. Others, believing that gardening is not going to be all that fun or relaxing after all, turn to seemingly less complicated pastimes, like jogging or tossing horseshoes.

And, so, fueled by a love of horticulture—and Meleah's aversion to sports (Jeff likes football)—we decided to write this book. In these pages we offer our fact-based thoughts on the usefulness and worthiness of some of the dos, don'ts, and shoulds we hear most often. It was no easy task to whittle these bits of advice down. You may not find every piece of advice you're wondering about, but we do hope to clear up a good share of the confusion that's out there.

As we dissected these various pieces of advice—good or bad—one of our objectives was to explain the reasoning and the research behind them. How often do you just do something in the garden because you have heard it's what you *should* do? Water deeply and infrequently, for example, or stop fertilizing in late summer. Why? It's important that gardeners know why we should do some things, but should not do others.

This book covers a wide range of subjects, such as soil, mulch, water, trees and shrubs, and vegetables. Each chapter is divided into three sections by the quality of the advice: Good Advice, Advice That's Debatable, and Advice That's Just Wrong. This way, if you just want to check out some good advice, you can quickly flip to that chapter and read that first section, saving the rest for another time when you might be in the mood for a good debate—or a laugh. By reading the Advice That's Debatable sections, you will start to understand some of the nuances of gardening. These sections help make clear how well-intentioned gardening advice can be a little off the mark. A case in point is tilling. If you type the word *tilling* into Google, you'll find just as many vehement supporters of the practice as detractors. Who's right? Can both sides be right? And if they are both right, how does that work? You will find the answer in the chapter on soils. In the last section, Advice That's Just Wrong, you'll laugh, you'll cry, and you may even shake your fist in the air. But we bet you'll find some surprises no matter how long you have been gardening.

For each piece of advice, we explain the practice itself before going into what will happen if you follow the advice, how to do it, and, finally, our bottom-line opinion or recommendation. As you read our thoughts on all of this advice, keep in mind that none of it is written in stone, which is for many people what makes gardening so attractive and fun. What we offer is the best information we can give based on current science, research, and, in most cases, experience—ours and that of our fellow gardeners. Of course, science is always changing, and one person's experience often differs from another's. So if you want to try something we are not too keen on, do it. Our mission here is not to stop you from doing certain things in the garden, but to provide good information so you can make your own decisions.

One last thing to know about the book is that even we do not always agree. Jeff is a scientist who is fascinated by the science of growing plants, whereas Meleah is more interested in actually growing the plants. We brought both of our perspectives to the project, and the finished book is the result of disagreements, compromises, and rewrites on both of our parts. Knowledgeable people debate gardening advice all the time, and knowing how to parse out that debate can be very helpful when you are trying to decide what advice you want to follow and what you can safely disregard.

As with Jeff's previous books on garden remedies and organic gardening, our goal here is to give you the information you need to make the best choices for you and your garden. We hope you will use this book as a reference the next time you hear some bit of advice and wonder how important it is to follow and what the ramifications will be if you don't. Does deadheading increase bloom? Is spring the only safe time to divide plants? Will overhead watering really result in a garden full of mildew and disease? What is full sun, and will you lose plants that require it but don't get it? Honestly, it can make your head spin. Well, never fear. Read on, learn more, and then figure out what you want to do in your own garden.

1

SOIL

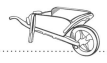 # In This Chapter

Good Advice

- Add organic material to all garden soil before planting

- Create an environment favorable to earthworms

- Use vermicompost to improve garden soil

- Always get a soil test before planting

- Add lime to soil to raise pH

- Add sulfur to soil to lower pH

- Stop fertilizing during very hot weather to reduce plant stress

Advice That's Debatable

- Change soil in flower boxes and other containers every season

- Use urine as a fertilizer

- Always fertilize in the spring

- Use peat moss for improving soil drainage

- Fertilize perennials and shrubs every year

- Till vegetable gardens every year

- Use organic fertilizers rather than synthetic fertilizers

- Use ammonium to green a plant and nitrate to grow it

- Add packets of bacteria to soil when you plant beans

- Use mycorrhizae to promote healthy plant growth

Advice That's Just Wrong

- Always use a balanced fertilizer

- Use pine needles to make soil more acidic

- Apply compost tea to help enrich soil

- Add sand to clay soil to improve drainage

Perhaps the most misunderstood ingredient needed for growing plants is the soil. And the reason is easy to see—or not to see, as the case may be. Because we cannot peer into the soil to tell exactly what's going on in there, it seems mysterious and unknown. We know that our plants' roots are down there somewhere mining about for water and food, but we are never sure exactly where they are unless we dig them up, and that is usually not a good idea.

Within the soil there are all kinds of things that together determine whether plants get what they need, such as earthworms, microbes, carbon, and silicon. Yet there is no shortage of gurus telling us that we need to add one thing or another to our soil to make it work better for our plants. Things like compost tea, mycorrhizae, worm castings, and compost. As you may already realize, some of this stuff works wonders, and some is hooey.

The truth is, in most cases, garden soils are appropriate for a wide range of plants without any work whatsoever besides a little bit of digging or tilling. Despite this, the recommendations continue. We are told that plants need earthworms, microbes, and fertilizers to make soil rich and welcoming. We are told that organic material, like compost, is a virtual necessity to make our earth appropriate for our crops. Meanwhile, off in a greenhouse or lab somewhere, a tomato plant with its roots immersed in sterile water and fertilizer is growing faster than any tomato plant that you have ever grown. It defies reason. If we don't always need soil amendments, how do we reasonably decide when we do?

Plants need to get three things from the medium that their roots live in: water, nutrients, and oxygen. In a hydroponic tub, all of these things are made available to the plant by humans. Water is abundant, the right amounts of nutrients are available, and oxygen is bubbled through the water to make sure plants roots can respire (believe it or not, plants breathe just like animals do). It's a little bit like an IV injected into a person to provide him with the fluids and nutrition he needs without actually requiring that the person eat. In a more natural situation, a properly functioning soil relies on the interconnections between minerals, organic material, microbes, plants, and animals. These interconnections are the key to having healthy plants.

The quality of a soil can be judged by its ability to hold and make water, nutrients, and oxygen available to plant roots. There are things we can do to make soil more effective at doing what it's supposed to do. The best soils for plant roots are well drained and moist and have an appropriate amount of nutrients and the microorganisms that process these nutrients so that plants can use them. A good soil should not be so full of nutrients that plants are poisoned by them.

good advice Add organic material to all garden soil before planting

Organic material is good for your garden. No matter what you are growing, you should seriously consider adding some. If you have ever looked at beautiful, black topsoil you know what a soil rich in organic material looks like. This soil is built up with organic matter from years and years of natural cycles of plant growth, death, and regrowth. It's true that there are some gardens that do not *need* more organic material. Still, it is not likely to harm them if you add some.

Organic material is primarily made up of carbon, and the two most common ways to increase carbon in your garden is to add compost or mulch. If you are fertilizing with manure, that will increase the organic material concentration as well. So will fertilizing with almost any fertilizer that includes plant materials, such as corn gluten meal or alfalfa meal. These organic fertilizers are based on dead and ground plants, which naturally contain a lot of carbon.

WHAT HAPPENS IF YOU ADD ORGANIC MATERIAL

Soil is made up of a great number of particles that are basically rocks, although they are much smaller than what we usually think of when we think of rocks. If these rocks are small, we call them sand. If they are really, really small, we call them clay. Sand is well aerated and does a good job of letting oxygen into the soil, but it does a poor job of holding on to water and nutrients that the plant can use. Because clay is made of smaller particles, it has properties that make it perform much differently than sand. Clay does not drain well, but it does hold water and nutrients very effectively.

So what's a gardener to do? Each type of soil has its advantages, but neither seems ideal. The answer is to add organic matter. Organic matter drains well and holds water and nutrients effectively. The best part is you do not need soil to be composed of 100 percent organic matter. In fact, just 5 percent organic matter in soil can work wonders.

If you never add organic material to soil, it will contain less over time as the material is lost to erosion. Such erosion is made worse if you till because tilling disturbs the soil's structure. Eventually, you will have a soil with very low organic matter, similar to what is found today on farmland across the country (which often has organic matter content as low as 1 or 2 percent). It might take many years for this to occur, but it will happen eventually. When the amount of organic material decreases in sandy soil, the soil holds less and less water. Clay, on the other hand, will hold more. Both will also hold fewer nutrients without organic material, and less oxygen will be able to get to plant roots, particularly in clay soils. In other words, without adequate organic material, soil will become much less able to support plants on its own and more reliant on irrigation and fertilizer provided by you, the gardener.

HOW TO DO IT

Organic matter can come from grass clippings, compost, mulches, and even weeds. Anything that was once living decays and goes back into the soil, returning organic material to it. That said, a few ways are more popular than others to get this valuable substance to where it can do the most good.

One of the easiest ways to boost organic matter is to turn 2 or 3 inches of compost into the top 6 inches or so of soil when starting a new garden. For established perennial gardens and other gardens that are not tilled, compost can be added atop the soil to a depth of about 2 or 3 inches, followed by about 2 or 3 inches of organic mulch, such as straw or wood chips, to control any weeds that may start to grow. Ideally this should be done every year, but even every two or three years will do soil a world of good.

Adding organic mulch alone to a depth of 2 to 6 inches each year will help supply organic material to the soil as it decomposes (we talk more about organic mulches in the chapter on mulch). Because all of these methods build up the organic matter in the soil, you have to worry less about watering. Regular applications of organic matter may even provide enough nutrients to make additional fertilizer unnecessary, too.

━● the real dirt *Whether you make your own compost in the backyard, buy it in bags, or have it delivered by the truckload, it is without doubt the most important soil amendment you can add to the garden.* Even if you start with the worst dirt imaginable, turning compost into the top few inches of soil each year (or even adding it as a topdressing that you do not turn in) will significantly change the quality of soil in just a few seasons. If you cannot manage to add compost to the whole garden in one year, just do a small part of it. Every bit helps.

good advice Create an environment favorable to earthworms

Earthworms have long been considered one of the most valuable creatures to take up residence in garden soil. But are they really that important? And what do they do that is so great?

The earthworms in our gardens and lawns are only a few of the approximately 6000 species of earthworms worldwide. These species can live in a variety of environments from old logs to mud. In fact, earthworms are present in almost all types of soil, but the healthier the soil, the greater their numbers will be. That is because healthy soil offers everything they need to thrive: air, moisture, and decaying matter to feed on.

All this feasting on decaying matter is great for garden soil, but it can be hard on forests, where earthworms are invaders rather than native inhabitants. Pushed out by glaciers during the last ice age, forestland in northern North America developed without earthworms until European settlers arrived, bringing earthworms along with them. Add to that all the earthworms used as bait that are tossed into forests by tired fisherman heading home, and you have got a full-fledged earthworm population explosion on your hands.

At the University of Minnesota–Duluth, Cindy Hale, a professor of forest ecology, has been working with other researchers to study the effect of earthworms on the state's forests. They found that in the absence of earthworms, decomposition of organic material on the forest floor happened slowly, controlled mostly by fungi and bacteria. This was great for certain plants growing in the understory such as tree seedlings, but not as great for others. With the introduction of earthworms, decomposition in the understory has been accelerated, leading to a reduction in the number of many native plant species normally found in hardwood forests.

WHAT HAPPENS IF YOU ENCOURAGE EARTHWORMS

In the garden, earthworms do all kinds of good things for the soil and the plants growing in it. As they move about, earthworms create tunnels that go deep into the ground, helping to break up compacted soil and allowing air and water to circulate more freely. They use their tunnels as burrows into which they pull all kinds of organic matter they find on the surface: pine needles, grass clippings, decaying leaves, and bits of wood. (The next time you dig up a chunk of healthy soil, look closely to see these tunnels.) Earthworms eat these things, along with bits of soil, and excrete castings (which you probably already figured is a gentler term for poop). The castings concentrate and make nutrients more available to plants.

HOW TO DO IT

By simply enriching soil with organic matter, you will encourage earthworms to live in your garden. Add compost and topdress with shredded leaves or grass clippings. Earthworms also like mulch quite a bit. This is a better strategy for increasing earthworm populations than going out and buying earthworms to bring home. Generally, earthworms do not like to be moved from place to place, and they often do not survive the transition from one site to another. If you do not have any earthworms despite your best efforts, get a soil test to make sure your garden does not have any heavy metals or other toxins and that the salt level is low.

━● the real dirt *Earthworms are nature's tilling machines. They do a great job of making nutrients, air, and water available to plants.* If you see a fair number of them in your garden, you can be pretty sure the soil is healthy enough for plants to thrive. Remember, though, earthworms are not a substitute for proper watering and nutrition.

good advice Use vermicompost to improve garden soil

Continuing on with our earthworm theme, let's take a look at vermicomposting. Gardening trends are always changing, and this practice is one of the hip things these days. Unlike the usual compost bin or pile in the backyard that includes yard waste, food, grass clippings, and other organic material, vermicompost bins (more commonly known as worm bins) are just for food scraps—and lots of wriggling worms. The earthworms eat up the scraps and leave behind castings that look a lot like dark, moist coffee grounds. These castings (vermicompost) are used in the garden to improve soil structure, attract more earthworms to the surface, improve water-holding capacity, and nourish plants by making soil nutrients more available. Earthworms are able to produce their own weight in castings in about 24 hours.

While the word *castings* is usually used interchangeably with the term *vermicompost* (as we are doing here), the two do not mean exactly the same thing. Castings are 100 percent worm poop, but vermicompost is a mixture of castings and decomposing organic matter. Dark, rich, and crumbly, vermicompost looks a lot like really good compost, and it is— only it's better because the nutrient content is usually higher than regular compost.

WHAT HAPPENS IF YOU USE VERMICOMPOST

Having a worm bin can be a fun and rewarding experience, but vermicomposting is also an inexpensive and very good way to boost the health of soil. You can skip all the work of having a worm bin and just buy castings to use in the garden, although they can be fairly expensive. But it's not a big deal if you do not use worm compost. Your plants will still be absolutely fine, as long as you are already taking care of the soil.

HOW TO DO IT

If you'd like to try vermicomposting, the first thing you need is a composting bin, and we recommend purchasing one. To gain insight into which kind might be best for you, take a look at Amy Stewart's book, *The Earth Moved: On the Remarkable Achievements of Earthworms*. It is an excellent resource for anyone looking to get into vermicomposting and very hard to put down once you start reading.

Next, you need to get some earthworms. It seems reasonable to think that earthworms are earthworms, but they are not. Various species do different things in the soil, and some cannot survive the confines of a compost bin. The earthworms you want are called red wigglers. If you want them to survive, they cannot be exposed to extreme heat or cold, which may mean the bin is kept indoors during some parts of the year, depending on where you live and if you share a home with open-minded people.

Steps to set up a worm bin are not difficult, but there is too much detail to go into here. You can find all of this information in Stewart's book, and Mary Appelhof's *Worms Eat My Garbage: How to Set Up and Maintain a Worm Composting System* is helpful, too. There are also good resources online, such as a publication from the University of Nebraska–Lincoln (available at http://lancaster.unl.edu/pest/resources/vermicompost107.shtml).

—▶ the real dirt *Other than the fact that you might not want to tend a box of earthworms, there's no reason not to try vermicomposting.*

good advice Always get a soil test before planting

Want to make an extension agent cringe? Just tell her that you wouldn't dream of wasting your money on a soil test. For academics working with plants, soil tests are treated as a necessity. Call up the local extension service to find out what went wrong with something in your garden and one of the first three questions will be, "Have you had a soil test done?" Soil tests are performed by many different institutions, both private and public, but most people use their extension services, which are based out of state universities.

The information provided by soil tests varies a bit depending on who examines it. Results will usually include the pH and the amounts of organic matter, salts, phosphorus, and potassium in the soil. Amounts of other elements, such as calcium, iron, manganese, and sodium, may also be included. Most testing services also offer the option of checking for heavy metals if you think those may be problems. Certainly all of this information is nice to have, but is it necessary?

WHAT HAPPENS IF YOU GET A SOIL TEST

By testing the soil where you want to plant vegetables, perennials, or trees, you are ensuring that conditions are right for them to flourish. Soil tests also provide information on what's wrong with a soil. Extension services receive more soil samples from people who have already planted something and are wondering why their plants are failing than from people who are testing their soil before planting.

Lots of people get along just fine without soil tests. If you plant a vegetable or perennial garden every year and you've done just fine without a soil test, why should you start now? That's a tough question to answer. Lots of gardeners at all experience levels do not get soil tests. If they have certain plants that do not seem to be doing well, they dig them up and give them away and find something that works. But for those who want to grow something in particular, like blueberries or azaleas, a soil test can be an important part of understanding how the soil might need to be amended to support these plants.

HOW TO DO IT

Getting a soil test is simple. You could test soil yourself with little kits you can buy at a garden center, but usually it is not much more expensive to send a soil sample to your state extension service and have them conduct the soil test for you. The extension service will provide general instructions for how to collect samples, but a few pointers include combining three or four samples from throughout the plot of land that you want tested to get an average. Also, make sure the soil that is sampled where the roots are living. In other words, don't just grab the very top layer of soil; instead brush away the top layer and get some from about 1 or 2 inches below the soil surface.

━➧ the real dirt *Soil testing is not necessary in the strictest sense, but it can be extremely helpful for determining soil health and what additives might be needed for growing a certain plant, like blueberries, which need to live in an acid soil.* A soil test can also be a helpful diagnostic tool if plants are struggling and there is no clear reason why.

good advice Add lime to soil to raise pH

This piece of advice is strongly related to the last. Plenty of old farmers and gardeners will tell you that you have got to add lime to your garden. Lime is a soil amendment, usually sold as a powder, that is composed primarily of calcium carbonate, though other elements could be present as well. The need for lime is largely dependent on the pH of the soil, and you will not know the pH of the soil unless you do a soil test.

There are many different types of lime, but the one that is most appropriate for most situations is dolomitic lime, which includes both calcium and magnesium. An application of lime can control pH for two or even three years, depending on the soil type.

Preferences for low and high pH

TOLERANT OF HIGHER pH (7–8)

TREES	SPECIES
American elm	*Ulmus americana*
Black locust	*Robinia pseudoacacia*
Bur oak	*Quercus macrocarpa*
English oak	*Quercus robur*
European mountain ash	*Sorbus aucuparia*
Flowering plum	*Prunus triloba*
Honeylocust	*Gleditsia triacanthos*
Pecan	*Carya illinoinensis*
SHRUBS	**SPECIES**
Alpine rhododendron	*Rhododendron hirsutum*
Buttonbush	*Cephalanthus occidentalis*
Chinese hibiscus	*Hibiscus rosa-sinensis*
Common lilac	*Syringa vulgaris*
Cotoneaster	*Cotoneaster tomentosus*
Firethorn	*Pyracantha coccinea*
Lemoine deutzia	*Deutzia ×lemoinei*
Oleander	*Nerium oleander*
Oregon grapeholly	*Mahonia aquifolium*
Peegee hydrangea	*Hydrangea paniculata*
Spirea	*Spiraea ×vanhouttei*
Sweet mock orange	*Philadelphus coronarius*
Tatarian honeysuckle	*Lonicera tatarica*
VEGETABLES	**SPECIES**
Brussels sprouts	*Brassica oleracea* var. *gemmifera*
Cabbage	*Brassica oleracea*
Chinese cabbage	*Brassica rapa*
Kidney beans	*Phaseolus vulgaris*
Mustard	*Brassica juncea*
Okra	*Abelmoschus esculentus*
Parsley	*Petroselinum crispum*
Peppers	*Capsicum* spp.
Pumpkins	*Cucurbita* spp.
Sweet corn	*Zea mays*
Turnips	*Brassica rapa*
Yams	*Dioscorea* spp.

TOLERANT OF LOWER pH (4.5–5.5)

TREES	SPECIES
American beech	*Fagus grandifolia*
American elm	*Ulmus americana*
American white birch	*Betula pubescens*
Cherry birch	*Betula lenta*
Flowering dogwood	*Cornus florida*
Pin oak	*Quercus palustris*
Plumleaf crab apple	*Malus prunifolia*
Red oak	*Quercus rubra*
Showy crab apple	*Malus floribunda*

SHRUBS	SPECIES
Blueberry	*Vaccinium corymbosum*
Carolina azalea	*Rhododendron carolinianum*
Doublefile viburnum	*Viburnum plicatum*
Japanese holly	*Ilex crenata*
Pink azalea	*Rhododendron periclymenoides*
Scotch heather	*Calluna vulgaris*
Staghorn sumac	*Rhus typhina*

VEGETABLES	SPECIES
Carrot	*Daucus carota*
Chives	*Allium schoenoprasum*
Eggplant	*Solanum melongena*
Garlic	*Allium sativum*
Potato	*Solanum tuberosum*
Sweetpotato	*Ipomoea batatas*
Tomato	*Solanum lycopersicum*
Turnip	*Brassica rapa* var. *rapa*
Watermelon	*Citrullus lanatus*

Liming is popular because certain plants, particularly legumes like soybeans, grow better at higher pH values. Actually, it is not the plants that prefer the high pH, but the bacteria that live symbiotically with them and collect nitrogen from the air, which the plants then use.

WHAT HAPPENS IF YOU ADD LIME

Lime will raise the pH of soil. If that is your goal, lime is the way to go. While most garden plants can handle a wide range of pH values, vegetables usually do best when the pH is between 6 and 7. When we say a plant is not able to handle a high pH, what that really means is that the plant cannot take up certain nutrients that it needs, usually iron and manganese, when the pH is too high.

Soil with a high pH that goes uncorrected will be fine for some plants. Most plants can handle soil pH levels between 5.5 and 7. The accompanying table shows some plants that do well at higher pH (7–8) and lower pH (4.5–5.5). Honeylocust and kidney beans, for example, have no problems in soil with a high pH. However, once the pH gets above 7.5 or so, most needle evergreens and many other plants start to suffer. pH levels above 6.5 are hard on acid-loving plants like azaleas and blueberries.

HOW TO DO IT

Although adding lime to a garden to increase pH is a good idea, in general it is not a good idea to apply lime without a solid understanding of the soil you have. We recommend a soil test before adding any lime. For those who are more adventurous than wise, the table offers some approximations for the amount of limestone to apply to increase pH to 6.5 based on the type of soil and the amount of pH change that you are seeking. Be aware, too, that if you add too much lime you will have a long wait, anywhere from one to three years, until the pH of the soil drops back down again.

Adding limestone to increase pH

SOIL TEXTURE	CURRENT PH			
	4.5–4.9	5.0–5.4	5.5–5.9	6.0–6.4
	Approximate pounds of limestone per 100 square feet of soil to increase pH to 6.5			
Sandy	11	9	7	2
Medium	18	14	11	7
Clay	27	23	18	11

━▶ the real dirt *Dolomitic lime is a very handy addition to acidic soils, and we have used it for certain plants to get them to grow well.* However, we have also been guilty of thoughtless, overaggressive lime application once or twice, severely stunting plants. And so, while we recommend lime if you need it, make sure you need it before you add it.

good advice Add sulfur to soil to lower pH

What if you need to lower soil pH? If after a soil test you find out that the pH of the soil is very alkaline, you can add a couple of different things to bring the pH down. Sulfur is the most common—and usually the least expensive and best—addition to lower pH. Sulfur has an acidic reaction with soil, which leads to a lower pH. This change in pH can last as long as two years and potentially longer.

WHAT HAPPENS IF YOU ADD SULFUR

Sulfur is the least expensive, easiest, and longest-lasting way to decrease the pH of soil. So it's an obvious choice if the soil pH is high. If you are trying to grow plants that prefer a low pH but you don't want to use sulfur because of the cost or because it takes a while to work and needs to be applied every few years, there are other options. First, you could grow different plants that prefer higher pH. You could also use an expensive acid-injection system to irrigate the land with slightly acidic water or use certain fertilizers, such as ammonium sulfate, which have an acidifying effect on the soil. Additionally, iron sulfate or a lot of peat can be incorporated into the ground. But, while all of these options have their place, sulfur is still the best answer in most circumstances.

HOW TO DO IT

After getting the results of a soil test, you can alter the pH of soil by one unit by adding ½ pound of sulfur (or 2½ pounds of iron sulfate) per 100 square feet of sandy soil, 1 pound of sulfur (or 5 pounds of iron sulfate) per 100 square feet of medium-textured soil, or 1½ pounds of sulfur (or 7 pounds of iron sulfate) per 100 square feet of clay soil. It will take a few weeks for the sulfur to work, so be patient.

—● **the real dirt** *The best thing to do when your soil has a high pH is to choose plants that can handle it.* If that is not an option or if you want to grow certain acid-loving plants, adding sulfur is the best way to lower soil pH in most situations.

....

good advice Stop fertilizing during very hot weather to reduce plant stress

During hot weather plants are under stress, especially if it has been dry recently. Plants can cool themselves naturally by transpiring, which is basically the plant equivalent of sweating. Water evaporating from the leaf surface helps the plant keep its temperature down. But if there is not much water available, transpiration does not work very well. Fertilizers are composed of salts, which make water even less available to plants. Even organic fertilizers make the soil somewhat salty and inhibit the uptake of water by plants.

Additional fertilizer can stimulate extra growth, thus compounding the water availability issue. This growth will be tender and need plenty of water, which, if there is a drought, will not be forthcoming. That makes fertilizing plants when it's hot one of the worst things you can do.

WHAT HAPPENS IF YOU STOP FERTILIZING DURING VERY HOT WEATHER

Plants will do better if you do not fertilize in hot weather. Most fertilizing should take place in the spring or autumn, when it is cooler and there's usually plenty of water around. If you decide to fertilize when it's hot, you will be putting a great deal of stress on your plants. Maybe they will be able to handle it and maybe they won't. Why take the risk?

HOW TO DO IT

Do all fertilizing in relatively cool conditions when you can water the fertilizer into the soil or potting mix. Though different plants have different ideas about what cool is, temperatures of 60° to 75°F (16° to 24°C) can be considered good for fertilizing. Many plants have met their demise because a gardener thought that the best cure for whatever their problem was would be pumping in more fertilizer. Do not make that mistake, especially in hot weather. Too much fertilizer is worse than too little. Be careful!

—● the real dirt *Stay away from fertilizing in hot conditions.* Spring and autumn are good times to fertilize for a lot of reasons, not the least of which is that you are less likely to hurt the plants you are trying to help if you fertilize during these seasons.

advice that's debatable Change
soil in flower boxes and other containers every season

When was the last time you cleaned the soil out of your flower boxes? Last year? Two years ago? Can't remember? Flower boxes are filled with material intended to keep the plants' roots happy. Though people usually call this material soil, it actually rarely contains soil. Most soils are not appropriate for use in containers because they hold too much water and not enough air. In other words, soil usually has very poor drainage when stuffed into a container. Instead, containers should be filled with potting soil, which is also called potting medium or artificial medium.

This potting soil usually includes organic materials, such as peat, coconut coir (basically the shell of the coconut), and composted pine bark, along with inorganic materials such as sand, perlite, and vermiculite. (Perlite is a type of volcanic glass that is extremely porous and lightweight. Vermiculite, which is also porous and lightweight, is made of expanded mica, a type of rock.) These types of mixtures are better for use in containers because they hold quite a bit of water and nutrients, are reasonably light so that the container will not be too heavy, and allow air to infiltrate the container to keep roots from drowning. The question is, with such nicely tailored media, why would we ever need to replace it?

WHAT HAPPENS IF YOU CHANGE POTTING SOIL YEARLY

Potting soil is expensive, so it's no wonder there is a lot of debate about whether it should be discarded after each season's use. If you follow this advice and start fresh with new potting soil every year, you will likely lose fewer plants to soil-borne diseases because you will eliminate the possibility of pathogens lingering from season to season. Another reason for changing out soil regularly is that vermiculite and perlite become compacted over time, so a container's drainage can suffer.

While it is not a bad idea to change potting soil in containers every year or so, the repercussions from not changing it will likely be quite mild. Yes, drainage will suffer over time. But the plant may well be able to adapt to these changes in drainage and not be badly affected. However, if a disease gets into a container, you could be in trouble. Not changing potting soil will allow the disease to fester and will eventually force you to either change the medium or abandon the pot. It comes down to deciding whether the risk is worth it or whether you'd rather spend the extra money and get fresh potting soil each year.

We agree with the growing consensus that discarding potting soil annually is probably overkill for most home gardeners. But it could be a good choice for gardeners who are growing cherished seed or a rare or exotic plant and would hate to lose it to disease or other problems. If you do reuse potting soil, it's a good idea to add some new compost from year to year to improve the soil structure and provide added nutrients. If you opt to get rid of the old potting soil, you can recycle it by adding it to an active compost heap, spread it somewhere else in the garden where you are less concerned about disease or just throw it away (which does seem like a waste of organic matter).

A BETTER WAY

Start with a potting soil of reasonable quality, because potting soil can be expensive. While you do not want the cheapest bag you can buy, which may be dry, clumpy, and buzzing with fungus gnats, you also don't have to run out and get the top-of-the line product that may make you think twice about container gardening. There are many different types of potting soil, but the best usually contain sphagnum peat along with composted pine bark and other ingredients. In general, it is best to stay away from potting media that includes a lot of vermiculite, as this can decrease the longevity of the media. You will also need to decide whether you want the soil to include fertilizer. In general, we do not recommend brands that include fertilizer, particularly quick-release fertilizer, because there's no way to control when and what plants are being fed.

Changing potting media is simple. Just dump out the old stuff and wash and scrub the container with water. If the container held diseased plants or you are starting plants from seed, before adding new potting soil use a mixture of 1 part bleach to 10 parts water mixture to sterilize the container. Then rinse the container with water to make sure all of the cleaning solution has been removed, as it can be dangerous to growing plants.

━● the real dirt *Fresh, sterile potting mix is a good choice for starting seed or for use with seeds or plants you have a sentimental attachment to or financial stake in.* Otherwise, reusing potting soil for a few years is a sensible, sustainable choice.

advice that's debatable Use urine as a fertilizer

Have you ever had neighbors who snuck out in the middle of the night to pee around their vegetable garden? Did you think they were slightly off the deep end? Turns out they might have been on to something. Urine does contain nutrients that are good for plants, most notably nitrogen in the form of urea. There are about 4 grams of nitrogen in every liter of urine. That is 4000 parts per million nitrogen, which is about ten times more nitrogen than what you might get with a typical liquid fertilizer application.

In other words, urine does contain stuff that will make plants grow. Unfortunately this nitrogen is so concentrated that, if enough is deposited in one small area, it could hurt a plant. While it may seem kind of gross to use urine as fertilizer, concerns that this practice could transmit disease are largely unfounded. This is because urine is relatively sterile when it leaves the body (unless there's a bladder infection), and because any organism that might be in the urine would not be able to survive once it was exposed to the outside environment.

That said, urine that is stored for any period of time in a cup, saucer, or other container can pick up nasty things. So, if you want to try using urine as a fertilizer, it's best to collect urine and put it directly on crops without any delay. Using urine in compost is also a very reasonable application.

The nitrogen will make the compost hotter because it will speed up the activity of the microbes decomposing the organic material.

WHAT HAPPENS IF YOU USE URINE

Using urine as a fertilizer, especially urine that is diluted, is an environmentally friendly idea, and it's not that disgusting to think of using urine to fertilize annuals or perennials even if you would not consider using it for vegetables.

However, peeing outside is a way to get yourself cited for lewd behavior in many residential neighborhoods, so you've got to be careful how you go about getting your pee to the garden. Besides, there are plenty of other ways to get nitrogen to your plants.

A BETTER WAY

The idea of using urine in compost is probably the least offensive and most reasonable of the techniques for using urine to benefit your garden. Placing urine in a compost pile is pretty straightforward: just collect your urine once a day or so and pour it onto the compost.

If you'd like to use urine directly on the soil around plants, dilute 1 part urine with 9 parts water and apply it once a week or so to non-edible plants.

the real dirt *There have been a few studies on the use of urine as a fertilizer, and most of these showed that it works fine.* Do not let urine sit around before using it, and we do not recommend trying it on edible crops—so don't add it to your compost pile if you use the compost for everything in your garden.

advice that's debatable Always fertilize in the spring

Over the years gardeners have developed practices that mimic the practices farmers use in their fields. One of these is fertilizing in the spring. On the surface, this makes sense. Because plants do most of their growing in the spring, it would seem that they need the most fertilizer during that time. In the past, many scientists recommended spring fertilization over autumn fertilization. But research is starting to show that fertilizing in the autumn can provide nutrients to plants just as well and maybe even better than spring fertilization.

WHAT HAPPENS IF YOU FERTILIZE IN THE SPRING

Fertilizing in the spring is beneficial to plants in some ways, but not in others. Fertilizers contain many different nutrients, and each of them works its way through the soil at a different speed. Nitrogen, for example, goes through the soil rapidly, whereas phosphorus and potassium move slowly.

If the soil in your garden needs more potassium and phosphorus, it's better to apply them in autumn, so the rain and snow of winter will drive these nutrients into the soil, where they will be available when plants need them come spring. Applying nitrogen too late in autumn, however, can be problematic. Over the late autumn and winter, much of the nitrogen may run right through the soil or off the top of it (particularly if the ground is frozen), so it never gets to the plants.

This is why early autumn fertilization is best for delivering nitrogen to many plants, including grass. Nutrients delivered at this time will be taken up and stored by the plant to be utilized in the spring during the first flush of growth. In fact, more nutrients will be taken from the plant's storage tissue to accomplish spring flush than will be taken from the ground. For annuals the equation is different. They take up and use nutrition during the same season, so spring fertilization makes more sense for them.

Skipping a spring fertilizer application is fine, even for annuals. If you are fertilizing at all, regardless of the timing, the fertilizer will make it into the soil and work for the plants. It just may not be at the levels you intended. Also, if you add compost fairly regularly, fertilization might not be necessary, because good compost takes care of most, or even all, of the nutrient requirements of plants.

A BETTER WAY

Apply fertilizer in the spring before the buds have broken on trees and the green tops of perennials have started to sprout. You want the fertilizer to be in the soil before the plants start growing, so they can use it when they need it. For annuals, wait to fertilize for a week or two after planting to avoid burning new roots.

If you are going to fertilize in the autumn instead, aim for early to mid autumn to give plants a chance to take up nutrients before winter comes and the soil freezes. Be careful about overfertilizing trees and shrubs in autumn, as this could stimulate new growth that could be damaged by an early frost.

Any fertilizer will have three numbers on the label, which stand for the percent of nitrogen, phosphorus (expressed as P_2O_5), and potassium (expressed as K_2O) in the fertilizer. We recommend staying away from high-phosphorus fertilizers because of their potential environmental impact. Instead, use a fertilizer with a first number that is higher than the second and third. For example, a 10–2–2 product has 10 percent nitrogen, 2 percent phosphorus, and 2 percent potassium. Generally, 1 to 2 pounds of nitrogen per 1000 square feet are recommended annually for ornamentals. Vegetable gardens will likely need more, with heavy feed-ers like corn and potatoes calling for as much as 4 pounds of nitrogen per 1000 square feet. If you are using a 10–2–2 fertilizer, then, you would need to apply 20 pounds of fertilizer in order to apply 2 pounds of nitrogen to a 1000-square-foot area.

—● **the real dirt** *Research is starting to show that au-tumn fertilization can be better than spring fertilization for perennial plants, but spring fertilization is certainly better for annuals and has worked for perennials for years.* Perhaps you could consider splitting fertilizer applications into two parts, half applied in spring and half in early autumn. Or you might consider using more compost so you don't have to use fertilizer at all.

advice that's debatable Use peat moss for improving soil drainage

Peat moss comes from plants that have been sealed away for hundreds, even thousands, of years in bogs. Over time, most of the nutrients have leached out of this moss, and what remains is a plant-based material that has a lot of good properties for growing plants.

Because it holds water and nutrients effectively, peat moss has long been known as a component of good, soilless media for containers. But some people will tell you that adding peat to a garden will improve drainage just like it does in containers. Although sometimes that's the case, its drainage qualities depend on how you apply it.

WHAT HAPPENS IF YOU USE PEAT MOSS

Peat moss is really just a bunch of plants that died a long time ago, so when you put peat into the ground you are adding organic matter. But you are also changing the pore spaces of the soil where you have placed it. If you add the peat to a large area, this is not a problem. But it can be a problem if you just add peat, for example, to a hole in which you plant a tree. The hole now has different pore spaces than the soil around it, so after a rain or a good watering, water will not be able to move easily from the hole into the surrounding soil. Because of this, you can easily end up with a pool in which the tree literally drowns.

There are plenty of ways to get organic matter into the soil without using peat moss, and some are better than others. Compost and mulch both add organic matter to soil, but they also have more nutritional content than peat, which is relatively nutrient free. Another important thing to understand about peat moss is that it's acidic. Wherever it is used, it will make the soil more acidic, too.

A BETTER WAY

In most cases, compost is a better choice than peat for improving drainage. Only gardens that are low in organic material and high in pH are really appropriate for an amendment of peat. If you do use peat, carefully follow the application directions on the package.

━●> the real dirt *Mixing peat into garden soil is not always a bad thing, but there are usually better choices, such as compost or shredded leaves, which have more nutritional value for plants.* The biggest problem with peat moss is that it may not be a good choice for the environment. There is an ongoing debate regarding the mining of bogs for peat moss, and boycotts of the product are often urged by those wishing to stop the destruction of the bogs.

advice that's debatable Fertilize perennials and shrubs every year

Every year the schedule is the same: plant, water, weed, and fertilize. But is this annual cycle really necessary? Or do the fertilizer companies have us bamboozled into thinking that we need to add nutrients annually when all we really need to do is leave well enough alone? Soils—especially soils with plenty of organic material—have the ability to hold nutrients for a long time, certainly longer than a single season.

If mulch is added to the soil every year, that will add some nutrients as it breaks down. Perennials and shrubs planted in the soil (rather than in containers, where a limited area for nutrients makes fertilizing a necessity) live for many years and develop extensive root systems that can mine for their own nutrients. So how much fertilizer do plants really need?

WHAT HAPPENS IF YOU FERTILIZE PERENNIALS EVERY YEAR

If you fertilize judiciously, you certainly will not hurt perennials and shrubs and you may be helping them to grow more quickly. But you need to ask yourself an important question: how quickly do you want your plants to grow? Trees, shrubs, and perennials have extensive root systems and, except in terrible conditions, are probably capable of fending for themselves. If you continue to feed them, they will get larger and larger—perhaps larger than you'd like them to be. If you avoid the fertilizer they may stay at a more manageable size.

Unless the soil around your plants is lacking a specific nutrient, not fertilizing will not harm them, providing that they are growing in good soil that is amended fairly routinely with organic matter. Most of the fertilizer we apply to our established trees, shrubs, and perennials is in excess of what the plants actually need. Another thing to keep in mind is if perennials are near grass that is fertilized, enough nutrients should leak through the layer of grass roots to satisfy these plants.

A BETTER WAY

Amend the soil around perennials, trees, and shrubs with plenty of organic material, and use mulch. If you do fertilize perennials, trees, and shrubs carefully follow the instructions on the package and stay closer to the low application levels than the high ones.

—● the real dirt *There are certainly situations in which poor soils demand more fertility or a particular plant requires a nutrient that is not present at a high enough concentration in the soil.* But, generally, we do not see a reason to fertilize perennials more than once every few years if the soil is good, meaning you add compost or other organic material on a regular basis. In fact, Jeff has not fertilized the perennials around his house for more than ten years now.

advice that's debatable Till vegetable gardens every year

Jethro Tull, who worked in England in the mid-eighteenth century, is perhaps the most famous agricultural scientist of all time. His work with a variety of agricultural equipment provided the basis for today's modern agriculture, with its use of tilling and planting in rows. Tilling is the practice of turning over all of the soil in a garden using a tiller or a similar piece of equipment. Tull believed that the roots of plants had tiny mouths that ate small pieces of soil and that pulverizing the soil would release its nutrients.

Though his reasoning was a bit off, tilling before planting generally encourages the growth of newly planted seedlings. Tilling mixes air into the soil, and oxygen encourages the activity of microbes that break down organic materials in the soil and release nutrients, especially nitrogen, in forms that can be absorbed by plants. Today, people often till their gardens every spring to prepare for another round of planting. Some till in the autumn, as well, mixing their spent crops with garden soil and clearing out old weeds.

WHAT HAPPENS IF YOU TILL EVERY YEAR

Tilling destroys weeds that might compete with vegetables and other plants. It also loosens the soil, reducing compaction and allowing air to circulate more freely. This sudden access to oxygen makes microorganisms become more active than they normally would be, and this in turn helps plants, particularly seedlings, to take up nutrients more effectively.

Unfortunately, this pulse of nutrients fades and ultimately leaves the soil poorer because the more active microbes consume so much organic material in the soil so quickly. Another downside to tilling is that by loosening the soil it allows organic matter to be more easily washed away, which, over the long term, is bad for your garden.

A BETTER WAY

Tilling is the easiest way to prepare soil to grow plants, but it is not necessarily the best way. Instead of tilling, many gardeners opt to simply amend the soil with organic matter to improve its structure and boost nutrients. Wood chips, bark, straw, and other materials can also be used as mulches on top of the soil to help keep weeds at bay, and as they decompose they will enrich the soil. Choosing to amend the soil with organic matter and mulch and pulling some weeds takes more foresight and work, but it will pay off in healthy soil in future years.

If you do choose to till regularly, it's best to do it only once per year, in the spring before planting. If you choose not to till as a regular practice, you still may want to consider doing so to get your garden started the first year or to incorporate some compost into the soil if it lacks organic material. Or, you can just dig the beds by hand and amend them as you go.

—▸the real dirt *Tilling repeatedly is bad for the long-term health of soil, but there's nothing wrong with tilling the soil when you are first planting a garden.*

advice that's debatable Use organic fertilizers rather than synthetic fertilizers

One hot-button topic that comes up for many gardeners is whether it is better to use organic fertilizers rather than synthetic ones. Organic fertilizers are composed of organic matter derived from plant and animal sources, such as corn gluten meal, fish emulsion, and composted manure. Synthetic fertilizers are chemically processed in ways that often use a great deal of energy and tax the earth and its ecosystems. However, some types of organic fertilizers, such as bat guano and rock phosphate, are harvested or mined in ways that are not ecologically friendly. Organic fertilizers offer many benefits, particularly their ability to release nutrients slowly rather than in quick bursts, as synthetic products often do (unless they are formulated for slow release). They may also contain micronutrients such as manganese, boron, or copper that synthetic fertilizers may not. Synthetic products often contain only nitrogen, phosphorus, and potassium, and perhaps a few other nutrients.

WHAT HAPPENS IF YOU USE ORGANIC FERTILIZERS

It is a common misconception among gardeners that plants respond better to organic fertilizers. In reality, both organic and synthetic fertilizers can provide the same nutrients to plants.

Gardeners choose organic fertilizers over synthetic ones for a lot of reasons. Some like the fact that organic fertilizers, such as alfalfa meal or fish emulsion, come from renewable resources. There is also the added benefit of increasing organic material when using organic rather than synthetic fertilizers. Organic fertilizers, however, usually have a lot of insoluble nutrients, which must be converted by soil microorganisms into soluble forms that can be absorbed by plants over several months or even years.

Though there are plenty of good reasons to use organic fertilizers, there are instances where synthetics are a better choice. The nitrogen, phosphorus, and potassium in synthetic fertilizers are usually almost immediately soluble in water, meaning that plants can take them up as soon as they are applied. This is good news for container-grown plants that are easily depleted of nutrients. Synthetic fertilizers also tend to be less expensive and more concentrated than organic fertilizers.

A BETTER WAY

If you are going to use organic fertilizers, then use ones that make sense ecologically. Compost and composted manure are at the top of the list, followed by things like alfalfa meal, corn gluten meal, seaweed extracts, and fish emulsion. Our table (page 45) provides a list of common organic fertilizers and typical amounts of the major nutrients found in them. When buying any organic fertilizer, be sure to check its list of ingredients, and be sure to follow the applications directions on the package.

━● the real dirt *Wisely chosen organic fertilizers are a great idea for the garden, but if you are on a budget, synthetics can do just as well. Aside from their sustainability issues, the biggest drawback to synthetic fertilizers is that they do not improve soil structure over time the way organic fertilizers do because they do not have any organic material in them.* Let's face it, synthetic fertilizers are much more convenient to use than organic fertilizers, so it's likely that gardeners are going to continue using them. When we do use synthetic fertilizers in the garden, though, it is wise to use them judiciously. In addition, it would also be a good idea to add compost and other organic matter as often as possible to ensure the health of the soil and all of the organisms and plants that dwell there.

....

advice that's debatable Use ammonium to green a plant and nitrate to grow it

Nitrogen is usually the nutrient that a plant uses the most of as it grows. In a farm setting, for example, as much as 500 pounds of nitrogen could be added per acre per year. Gardeners do not usually use as much, especially if we have been maintaining high levels of organic material in the soil through the use of compost and mulch. Still, we rely on nitrogen to help our plants grow.

There are two types of nitrogen that you can buy: ammonium, found in chemicals such as ammonium sulfate, and nitrate, found in chemicals such as calcium nitrate and potassium nitrate. You can even find these two chemicals together in the form of ammonium nitrate, which is used in many synthetic fertilizers. When buying a synthetic fertilizer you will find the amounts of ammonium and nitrate on the product's label. This can help you to decide which type of fertilizer is right for you. Old-time farmers used to say that ammonium greens a plant and nitrate grows it, which is basically true. But there's more to it than that.

WHAT HAPPENS IF YOU USE AMMONIUM OR NITRATE

Plants take up ammonium and nitrate somewhat differently. Ammonium is taken up more readily than nitrate, and plants respond by greening up faster. Nitrate is what ammonium turns into after soil microbes get done with it, so generally you will find more of this type of nitrogen in garden soil. Because nitrate is more prevalent in soils, it usually provides most of the nitrogen for plants' growth.

Plants generally use more nitrate than ammonium. Though ammonium is taken up more readily, high levels of this form of nitrogen can be toxic to plants. Under cold conditions ammonium is not readily changed into nitrate by soil microbes, so growers who are wary of ammonium damage during cool seasons will feed their plants with nitrate.

The idea that ammonium greens a plant and nitrate grows it is basically true, but it's not something that you need to worry about in most situations. It's okay if you do not select fertilizer based on the type of nitrogen it contains. Microbes in the soil will convert most of the ammonium to nitrate. What you do need to consider is using a nitrate-based fertilizer if you are going to fertilize in cool or even cold weather.

A BETTER WAY

Any type of nitrogen fertilizer that includes nitrate will have the word *nitrate* in the list of ingredients. Ammonium will be listed as either ammonium or urea. If you are fertilizing at temperatures below about 70°F (21°C), particularly if you are fertilizing containers, consider using a fertilizer in which the nitrogen is almost entirely nitrate. If temperatures are higher, it is safe to use ammonium (including urea).

—● **the real dirt** *In most cases, there's no need to differentiate between the two types of nitrogen. In cooler weather, however, it's best to use nitrate.*

....

advice that's debatable Add packets of bacteria to soil when you plant beans

Many garden centers sell little packets of bacteria for bean plants. These are a special type of bacteria that grow symbiotically with beans and other legumes. They take nitrogen out of the air and fix it so that it is available for the plants to use. These plants and their associated bacteria were the primary way that nitrogen found its way from the air to soils before synthetic nitrogen became available.

WHAT HAPPENS IF YOU ADD BACTERIA

These bacteria grow within nodules on the plants' roots, so they need to be in the soil where the beans are planted. In places where beans are commonly grown, gardeners can expect these bacteria to be there waiting to help beans gather nitrogen. Even in soils where beans or other legumes are not normally grown, if legumes are planted, the bacteria will eventually find their way to the plants' roots. However, adding bacteria from a packet can be helpful for young plants because it can give them a head start and make them more productive as the season progresses.

If you do not add the bacteria from these packets, your beans will be fine. We have both planted plenty of beans and other legumes without adding bacteria and ended up with good crops. In fact, when the soil is high in nitrogen, these bacteria may not even associate well with the plant.

A BETTER WAY

If you want to grow beans without using a lot of fertilizer in a plot where beans haven't been grown before, there are various ways to apply the bacteria to the plants. Directions on the packet do need to be followed, but, generally, if the bacteria are anywhere close to where the bean seeds are germinating, the plant will associate with the bacteria and a relationship will form. If the plot is being fertilized heavily or has had beans in it before, the packets of bacteria probably aren't needed.

—● **the real dirt** *Using bacteria is not necessary, but it can really help if beans or other legumes have not been growing in that region of your garden for a while.*

Major nutrients in common organic fertilizers

ORGANIC FERTILIZER	NITROGEN N (%)	PHOSPHORUS P (%)	POTASSIUM K (%)	NOTES
Alfalfa meal	3	1	2	Renewable fertilizer derived from alfalfa
Fish emulsion	5	2	2	Soluble liquid fertilizer, which can smell terrible
Bat guano	10	4	2	Harvested from caves, which may have ecological consequences to their flora and fauna
Milorganite	7	3	0.5	Made from human waste, but considered safe and relatively environmentally friendly
Cotton seed meal	6	1	1.5	Renewable fertilizer derived from cotton
Corn gluten meal	10	0	0	Renewable fertilizer derived from corn; also a pre-emergent herbicide
Seaweed (powdered kelp)	1	0	4	Renewable fertilizer derived from seaweed; rich in micronutrients
Cow manure	1	0.3	0.5	Use composted manure to reduce the danger of pathogens and weed seeds
Horse manure	1	0.3	0.5	Use composted manure to reduce the danger of pathogens and weed seeds
Chicken manure	25	15	10	Use composted manure to reduce the danger of pathogens
Rock phosphate	0	30	0	Derived from mines; not considered a renewable resource
Greensand	0	1	6	A mined resource
Soybean meal	7	2	1	Renewable fertilizer derived from soybeans
Blood meal	12	1	1	From dried blood, primarily cow
Bone meal	3	20	0	From ground bones, primarily cow
Feather meal	10	0	0	Feathers from chickens; very slow release of nitrogen
Fish meal	10	4	0	Derived from ground fish

advice that's debatable Use mycorrhizae to promote healthy plant growth

Some recommendations make sense in theory and, yet, when they are actually tested, they just don't work the way that they are supposed to. Mycorrhizae fit nicely into this camp. You can buy mycorrhizae at just about any garden center. These fungi will supposedly help plants gather nutrients and water, and in a normal forest setting that is indeed what they do. They accomplish this by connecting to the roots of the plant and essentially acting like surrogate roots. Some plants actually require this kind of mycorrhizal interaction to live. The problem with mycorrhizal products is that the fungi are frequently dead by the time you buy them. Even if they are not dead, the soil where you want to use them probably already has its own mycorrhizae and does not need an extra shot, so buying the fungi is a waste of money.

WHAT HAPPENS IF YOU USE MYCORRHIZAE

There is no question that naturally occurring mycorrhizae are beneficial. The only questions are whether you need to add them to soil and whether they are alive when you buy them. In some instances, adding mycorrhizae might be beneficial. For example, if you have inadvertently sterilized your soil recently by using a tremendous amount of synthetic fertilizers or pesticides or you have had some construction done that affected the soil, adding the fungi could be helpful. It's also reasonable to add mycorrhizae to soil that has been covered for a long period of time (at least a few weeks) with a layer of plastic, as you might do when you want to kill weeds and diseases. In most cases, though, it will not make a difference one way or the other if you do not add mycorrhizae.

A BETTER WAY

In a case where you suspect soil may have been sterilized and may need mycorrhizae, add mycorrhizae according to the instructions on the package. However, in most instances the best way to keep soil inoculated with mycorrhizae is simply to make sure that it is well fortified with organic materials such as compost and mulch.

→● the real dirt *While we do not believe that mycorrhizae are a scam, we do think that they are recommended far more often than they are needed.* When we have tested them at the University of Minnesota with otherwise healthy plants growing in reasonably healthy soils we have not found any benefit to their use.

advice that's just wrong Always use
a balanced fertilizer

An old-fashioned recommendation that some people still follow is using a balanced fertilizer, that is, one that contains roughly the same amount of nitrogen, phosphorus (expressed as P_2O_5), and potassium (expressed as K_2O). For example, 10–10–10 is considered a balanced fertilizer. The problem with balanced fertilizers is that they contain more phosphorus and potassium than what plants usually need. Usually a fertilizer with a smaller quantity of phosphorus and potassium is preferred, something with a ratio along the lines of 5–1–2 or 5–1–3. The big problem with balanced fertilizers is that nitrogen moves through the soil rapidly, whereas phosphorus and potassium stick around. Thus, after a few years of adding balanced fertilizers, even in soils that might be naturally low in these elements, you end up with high levels of phosphorus and potassium.

WHAT HAPPENS IF YOU USE A BALANCED FERTILIZER

In situations where a garden really does have low phosphorus and potassium, balanced fertilizers are a great idea. But without a soil test revealing these kinds of deficiencies, adding a balanced fertilizer does not make sense. Even if your soil is low in phosphorus and potassium, it is not a good idea to continue adding a balanced fertilizer over many years because these elements will not move rapidly out of the soil after they are applied. Thus, a balanced fertilizer should be used for two, three, or maybe four years, but no longer.

If you stay away from balanced fertilizers, you will avoid using phosphorus and potassium that your garden likely does not need. This can save you money and helps the environment, because excess nutrients can end up leaching or running off into local waterways and groundwater, where they may promote the growth of algae and reduce populations of fish and other aquatic creatures.

A BETTER WAY

Before choosing which fertilizer to use in your garden, have a soil test performed. Instead of a balanced fertilizer, you will likely need one with higher levels of nitrogen than of phosphorus and potassium.

━➧ the real dirt *Balanced fertilizers are rarely recommended these days. In general, it is best to stay away from them.*

advice that's just wrong Use pine needles to make soil more acidic

In a way, this advice provides two pieces of gardening information. The first is that gardeners can use pine needles (that is, the needles of evergreen trees) as mulch to acidify garden soil. The second is a cautionary tale that suggests pine mulch can reduce soil pH so much as to be harmful to plants.

Indeed, there are plenty of gardeners who use pine mulch successfully. Yet most of us have noticed that few plants grow underneath evergreen trees where the needles fall. What's going on? The problem has much more to do with the fact that the tree itself produces inhospitable, dry, dense shade underneath it than with the soil pH.

Research backs up this opinion. For example, David Zlesak, an assistant professor at the University of Wisconsin–River Falls, recently tested the area below various evergreens growing in a variety of soils. He found no differences between the pH directly below trees where needles built up and in nearby soil where there were no needles.

WHAT HAPPENS IF YOU USE PINE NEEDLES

We do not want to discourage you from using pine needles as mulch, because they can work well. But if you are looking for something to make soil more acidic, this is not the stuff.

A BETTER WAY

If you need to acidify the soil in your garden, adding sulfur will produce good results. To use pine needles as a mulch, spread them to a depth of 2 or 3 inches around the area you want to be free of weeds. Pine needles are good at blocking weeds, because they may contain chemicals that have the ability to prevent seeds from germinating. Therefore, it's usually best to avoid putting pine straw around seeds that you want to encourage.

━● the real dirt *Pine straw is a great mulch, but don't expect it to do anything to the pH of your soil.*

advice that's just wrong Apply compost tea to help enrich soil

Compost tea is a product that can be added to soil to, supposedly, make it rich and vibrant. This tea is made by steeping a nylon bag full of compost in a tub of water for a day or more. The tub often has air bubbled through it (producing something called aerated compost tea) and may or may not have additional sugars, such as molasses, added. The theory is that while the compost is soaking, good bacteria and fungi reproduce in the tub. The resulting tea is said to contain beneficial microbes as well as nutrients to help plants grow.

It is true that compost contains beneficial bacteria and fungi as well as nutrients. Proponents of compost tea say that the air bubbled through it stops bad bacteria from growing, so when you apply it you are giving soil a shot of good microbes. Unfortunately, researchers have been unable to demonstrate these benefits on any sort of a consistent basis. Worse yet, bad bacteria, such as *Escherichia coli,* can infest the tea, particularly when sugars like molasses are used.

WHAT HAPPENS IF YOU APPLY COMPOST TEA

If you think that your soil is deficient in beneficial microbes, we recommend skipping the compost tea and using compost instead. Compost, with its excellent ability to hold water, nutrients, and air, provides not only benefi-cial organisms, but also a place for them to live. Applying compost tea to a poor soil is a lot like dumping a whole bunch of people in the middle of the Sahara Desert without any supplies. They are not going to build a new civilization; they are going to die.

At the risk of sounding really grumpy about this whole compost tea thing, we would also like to point out another obvious problem. Even if you brewed the tea the same way every, time there is no way to control the amount of fungi and bacteria it contains. So why go to all that trouble when you can simply add compost and know you are adding something that will be ben-eficial? Besides, you will avoid the risk of poisoning yourself by pouring an *E. coli*–laced mixture on your vegetable garden.

A BETTER WAY
The way to improve soil is with compost and mulch. These will provide the microbe-friendly organic matter that good fungi and bacteria need.

━● the real dirt *Compost tea is an unproven concoction that has the potential to make people sick.* At this point, we do not recommend it and we strongly encourage people to think twice before even testing it.

....

advice that's just wrong Add sand to clay soil to improve drainage

Woe to the gardener who has to work with clay. It is composed of very fine bits of soil, so clay is extremely efficient at holding water. This makes it drain poorly and can create inhospitable conditions for a lot of plants and ongoing headaches for gardeners.

Add sand, some suggest, and the coarse material will make clay less water-logged. In truth, though, when sand and clay are mixed together they blend in such a way that they create a dense, heavy mess—one with a consistency akin to wet concrete. If you wanted to alter clay with sand, you'd need about a 1:1 mixture of clay soil to sand, and that's a whole lot of sand to haul in. Organic amendments are a better choice for making clay soil easier to work with. Even organic mulches, such as wood chips, will help because they will break down and slowly work their way into the soil.

WHAT HAPPENS IF YOU DON'T ADD SAND

Unless you have a lot of sand available and a strong back to move it all, don't consider it. Clay is not the end of the world. You just need to figure out what grows best in a soil that tends to be wetter, heavier, and stickier than other soil types.

A BETTER WAY

Rather than using sand, if you want to make clay soil a bit more workable, stick with organic material. Compost is the best choice, but using mulch is fine, too. If you do use mulch, apply it to a depth of 4 to 6 inches if you can, particularly around shrubs and larger perennials. As the mulch decomposes, it will add organic materials to the soil and improve drainage. Remember, though, there's only so much you can do with clay soil.

When you have clay soil, to some extent, you need to just accept it and work with it to the best of your ability. There are many plants that grow just fine in clay soil, such as columbine (Aquilegia canadensis), New England aster *(Aster novae-angliae),* purple coneflower *(Echinacea purpurea),* daylily *(Hemerocallis),* blazing star *(Liatris),* Russian sage *(Perovskia atriplicifolia),* stonecrop *(Sedum),* and common lilac *(Syringa vulgaris).*

━▶ **the real dirt** *Concrete is a poor medium to plant into, so do not add sand to clay.*

2
WATER

In This Chapter

Good Advice

∘ Water deeply and infrequently to encourage a strong root system

∘ Add water to the hole before planting a tree or large shrub

∘ Water only in the morning

∘ Check soil moisture before watering

Advice That's Debatable

∘ Avoid overhead watering to control plant diseases

∘ Water only young, newly planted trees

Advice That's Just Wrong

∘ Wait until plants wilt before watering

∘ Use gravel or rocks at the bottom of containers to improve drainage

There is nothing more fundamental to successful gardening than proper watering. When someone comes to us saying they are having trouble with a plant, the first question we ask is, "How often are you watering?" In our experience, overwatering is just as common as watering too little, and the symptoms of both are very similar. This makes it difficult to diagnose one or the other just by looking at a picture of the dead or dying plant.

The movement of water is critical because water affects every part of a plant. The path water takes along a plant's body from its roots to its leaves is long and arduous. Any break due to the soil drying out can have far-reaching consequences for its health. For example, plants that suffer severe water stress can continue to suffer for years afterward, growing more slowly than their unstressed counterparts and with a decreased ability to deal with problems such as insects and disease.

Plants do not just need water around their roots, they need air around their roots, too. Plants can live with their roots sitting in water, as they do in a hydroponic setup, only if the water has air bubbled through it. The bubbling keeps oxygen in the water and allows the plants' roots to respire (breathe), as they need to. Without oxygen, the roots will rot and die. This is why drainage is so important for plant roots. Good drainage ensures that water moves rapidly through soil or potting media, allowing air to fill the spaces that are vacated.

good advice Water deeply and infrequently to encourage a strong root system

This is true for most plants, and the reason is pretty straight-forward. Roots grow where there is water. If water is not available in a particular area, roots will not grow there. This means that if you're the type of gardener who sprinkles the surface of the ground every day, the roots of the plants will only grow in that upper layer of soil. Moisture in the top layers of soil is depleted more rapidly than in deeper layers due to heat and wind.

If you want roots to grow more deeply—and you do—you have to water for longer periods and less frequently to encourage them to grow down to greater depths. Once plants are rooted more deeply, they require less water because they are better able to take up water that lingers further down in the soil.

WHAT HAPPENS IF YOU WATER DEEPLY AND INFREQUENTLY

A deeper, stronger root system is good for the survival of grasses, annuals, perennials, vegetables, shrubs, and trees during lean times. Scant watering on a frequent basis creates shallow root systems that cannot sustain plants under tough conditions. Plants with shallow roots may look fine most of the time, but when a drought comes or you go on an extended vacation when there is not much rain, your plants will suddenly look terrible because that top layer of soil will not hold the water the plants need.

HOW TO DO IT

To get nice deep roots you need to water deeply just often enough to keep plants from showing any signs of water stress, such as wilting. How frequently? Well, obviously, plants have different watering needs. For example, grass needs to be watered more often than trees.

As a gardener, you will get to know the needs of your own plants, but we can provide some general guidelines for watering. In the heat of the summer, if no rain is on the horizon, grass should be watered once every two or three days. Shrubs and perennials need a good soaking every three or four days. Vegetables will usually need to be watered every two or three days at first, and then every three or four days as they get older, depending on how hot and dry it is and the quality of your soil. Young trees need water about once or twice a week, depending on when they were planted, their size, and how sandy the soil is.

Each time you water, soak the top 5 or 6 inches of soil, which adds up to about 1 to 1½ inches of water. Get a rain gauge or use an empty tin can to track how much water Mother Nature offers up, and then add the rest yourself.

The best way to test when the soil is wet to the correct depth is to push a spade or ruler down into the dirt to create a little gap. Then, stick your finger down in there to see whether the soil is moist. Another trick for young trees is to remember that each week they need about 5 to 7 gallons of water applied over the area known as their drip line, the ground beneath the tree's canopy.

—● the real dirt *Infrequent, deep watering allows plants to establish the deep, healthy root systems they need to handle dry periods and avoid drought stress.*

good advice Add water to the hole before planting a tree or large shrub

If there's been a lot of rain shortly before planting time, this is not necessary. But if it's been dry, adding water to the hole before planting is a really good idea. When you are planting a new tree or shrub, it's best for the roots to be moist right away. By filling the planting hole with water, you are ensuring that its roots will not start to dry out immediately after you throw on the last shovelful of soil.

Filling the hole with water also helps to ensure that there will not be air pockets, which could be detrimental to new roots. Air pockets can prevent roots from growing in the direction of the pocket and create problems as the soil in the planting hole settles. Some people prefer to make a slurry of the ground where they are planting, especially if they are planting their trees and shrubs bare root, a harvest technique where the tree or shrub comes without any soil or container mix around its roots.

WHAT HAPPENS IF YOU ADD WATER TO THE HOLE

Particularly during dry spells, adding water to the hole helps to provide moisture to tree and shrub roots at planting time. Otherwise, the tree could end up in soil that is not moist enough for its needs.

If the weather has been wet recently, it should not be a problem if you do not add water to the planting holes. Also, if you are careful to add water immediately after planting, it is not as important to add water while planting.

HOW TO DO IT

After you are finished digging the planting hole, before you plop in the plant, pour in a few gallons of water (you do not need to be precise). You just want to fill about a quarter of the depth of the hole. It is not necessary to let it drain before you plant the tree, although leaving the water in the hole while you plant may make a messy job even messier.

—➡ the real dirt *Adding water to a planting hole is a good idea.*

good advice Water only in the morning

It has long been said that morning is the best time to water, and there are some good reasons for that. First, there is usually less wind in the morning, so if you are watering with a hose or sprinkler the water is less likely to evaporate. Cooler temperatures in the morning also help to alleviate evaporation.

It is also said that morning watering helps decrease the chances of leaf burning due to water droplets creating a magnifying glass–like effect by focusing the sun onto leaf surfaces. Research conducted by Gabor Horvath and his colleagues at Eötvös University in Budapest, Hungary, showed that plants with hairs on their leaves, such as ferns, can collect water on those hairs. In the sun, the water on the hairs can have the same effect as a magnifying glass, causing leaves to burn. If the leaves are not hairy, though, it is not likely that any significant magnification of sunlight will occur.

WHAT HAPPENS IF YOU WATER ONLY IN THE MORNING

Morning has a lot of things going for it, not the least of which is that it's more environmentally friendly to water in the morning in terms of conserving water. However, if the only time that you can water is in the afternoon or evening, then you should water at those times. When a plant needs water, it needs water. Waiting until morning to give it that water may be a bad idea for the plant.

HOW TO DO IT

The best time for watering is right around sunrise, if you can manage it, when winds are at a minimum and it has not yet begun to get warm.

—● the real dirt *The morning is the best time to water, but don't wait until the morning to water if your plants are showing any signs of water stress.* Also, be aware that, in some cases, plants growing outdoors in containers may need to be watered two or even three times a day.

good advice Check soil moisture before watering

There is a common misconception among gardeners that when it comes to water, more is better. This is not the case. A lack of water and too much water cause many of the same problems for plants.

For example, both inhibit water from traveling up a plant's stem, which causes wilting and then browning of leaves—and possibly even death. Because the symptoms of over- and underwatering often look very much the same, gardeners may pour water on drowning plants because they think that wilting is due to underwatering when in reality the opposite is going on. That is why it's so important that gardeners always check soil moisture levels before watering.

WHAT HAPPENS IF YOU DON'T CHECK SOIL MOISTURE

If you want healthy plants, you *always* have to think about how much water you are providing them. Loving your plants too much is the surest way to send them to an early grave. Some plants handle being over- or underwatered better than others, but they will not be able to tolerate one or the other for long, so it's best to get watering under control from the start.

HOW TO DO IT

Making sure that plants are properly watered can be as easy or as complicated as you want to make it. The easiest and most reliable way to test soil moisture is to simply push your finger down into the soil to see whether it's dry, moist, or wet. Ideally, soil should feel moist but not wet.

Don't want to stick your finger into the soil around plant after plant? You can try clay worms or other clay doodads that can be inserted into the soil. When you water these gadgets turn from pink to red. As they dry out they go back to pink, letting you know it's time to water again. These work, although they are not as accurate as the finger test. You can also try electronic moisture sensors, which generally do a pretty nice job of telling you when the moisture level is correct.

Once a year or so, it is a good idea to remove container plants from their pots right before a watering to see whether the container is really drying out as much as you think it is. Often, we water container plants based on what the top of the soil looks like, not realizing that the middle and bottom of the container are holding a lot of water. Sometimes this problem is so severe that the roots are rotten at the base of the container.

If you see this, be sure to replace the media in the container and water less frequently. (If there are rotten roots you should consider getting rid of the plant, too. At the very least you should prune the rotten roots off.) As media ages, it gets compacted and impairs drainage, so it tends to stay wet and soggy longer. Repotting plants on a regular basis will help with this problem.

━● **the real dirt** *Don't just water your plants on a regular schedule.* Check the soil to be sure you really need to water before you do it.

advice that's debatable Avoid overhead watering to control plant diseases

There is no debating that overhead watering systems often lead to plant disease because wet leaves provide the ideal environment for spreading all kinds of disease pathogens. That is why drip systems, like soaker hoses, which wet the soil but keep leaves dry, are recommended. Overhead watering also tends to be terribly inefficient, with much of the water applied evaporating or landing on an area that does not need to be watered, such as a driveway. Still, soaker hoses can be a pain, especially if you have a large garden or several gardens. So a lot of gardeners just get out the overhead sprinkler and let it rip—preferably in the morning, so leaves have a chance to dry before the temperature drops at night. If you do this, as Meleah admits to doing, just be aware of the risks.

The other reason overhead watering is a debatable practice is because in the case of a very few diseases, such as powdery mildew, spraying affected leaves with water on a regular basis can actually help to control the problem. This sounds odd, we know, because people often wrongly associate powdery mildew with humidity and moisture, but powdery mildew is one of the few diseases whose spread is inhibited by water. Powdery mildew performs best when it's warm and humid, but without precipitation. Water on a leaf's surface will inhibit this disease.

WHAT HAPPENS IF YOU DON'T AVOID OVERHEAD WATERING

Overhead watering is not advised but it's a reality for many gardeners who don't have the time, patience, or drip system infrastructure to do it any other way. If you do not heed the warnings to avoid overhead watering, you may wind up having to fend off more diseases in your garden than you would have otherwise.

A BETTER WAY

Drip systems are a good alternative to overhead watering, and soaker hoses are the most popular choice. Some gardeners also water plants by hand with a watering can or spray nozzle attached to the hose. If you do this, use a gentle spray, direct the water around the base of the plant, and try to keep most of the moisture off the leaves. While this can be a very relaxing activity and allows you to get a good look at your plants, it is not very practical if you have a lot of gardens to care for.

━● the real dirt *We support the use of drip systems because overhead watering is inefficient and can lead to disease problems.* But we also understand that, for practical reasons, many gardeners use overhead sprinklers and take their chances with disease.

advice that's debatable Water only young, newly planted trees

When it's time to plant a new tree in your yard, the need for water is obvious. Not only are trees often visibly dry at planting time, tree roots are invariably damaged in the process of transplanting and need to be given extra care through their first year or so. As they are dug from the ground and put in containers or balled-and-burlapped, roots go through a lot before we ask them to spread out and do their job in a new spot. During that first season, before a new root system has been established, a young tree relies on moisture from its root ball to survive. That's why it is so important that trees are watered well at planting time and regularly for that first year.

Things are different for older, established trees. These trees have large, established root systems that canvas the soil all around the tree for moisture. Generally speaking, trees have survived for millions of years without people around to water them. So you don't *have* to water trees after they have been in the ground for two or three years. If you do, though, the water may help them grow and thrive because drought stress can make trees more susceptible to disease and pest problems.

WHAT HAPPENS IF YOU WATER YOUNG OR OLDER TREES

Young trees need water in order to become established in a new site. Because of their extensive root systems, older trees do not usually need to be watered unless there is a drought.

A BETTER WAY

Watering trees is a tricky business because they are so easily overwatered. When providing water to a young tree, consider the ground around the tree. If the soil is sandy, you should water more often than if you have clay, which tends to hold more moisture. Remember that overwatering is just as bad as not watering enough, so check the soil around the tree before reaching for the hose.

Water deeply and infrequently so the young tree establishes a reliable root system that can get it through dry times. As a rule, young trees need about 1½ inches of water each week. When you water, a soaker hose is usually the best piece of equipment to have. Just turn the hose on low and let it run for a couple of hours. If you are able, move the hose around beneath the tree's drip line as you water so the entire root zone gets moisture. This same system also will work for older, more established trees with larger drip lines, but you will need a longer length of soaker hose.

—• the real dirt *Young trees that were transplanted have compromised root systems, so they absolutely need to be watered for the first two to three years.* Larger trees, on the other hand, rarely need to be watered unless there is a drought.

advice that's just wrong Wait until plants wilt before watering

A plant's first response to drought stress—the one we should be responding to—is one we cannot see. Before they wilt, plants close off tiny pores on the undersides of their leaves. These pores are surrounded by guard cells, which block the pore when there is not much moisture available. But guard cells do not completely cover the pores or prevent loss of water, so plants have another more obvious response when they are parched: wilting. Wilting helps protect plants from losing even more water by causing leaves to fold around the pores. This reduces transpiration from the undersides of leaves and keeps temperatures down by limiting the amount of sunlight hitting leaf surfaces. You can probably see where this is going: by the time plants have started to wilt, they may have already suffered significantly.

WHAT HAPPENS IF YOU WAIT UNTIL PLANTS WILT

If you wait until plants wilt before you water them, they probably will not die under your care. But they will be small, gangly, and prone to leaf drop regardless of the season. If you monitor soil moisture using something other than wilt, such as your finger or some gadget, your plants will do much better.

A BETTER WAY

Rather than allowing your plants to wilt, use the methods to measure moisture described earlier in this chapter, such as testing the soil with your finger. Relying on wilting is a sure way to make plants suffer.

—• the real dirt *Letting a plant wilt once for a short time is not a problem, but letting it wilt regularly is a surefire way to stunt the plant and keep it from thriving.*

advice that's just wrong Use gravel or rocks at the bottom of containers to improve drainage

One of the oldest gardening recommendations dates back to the 1700s and comes from Phillip Miller's **The Gardeners' Dictionary.** It goes something like this: add gravel (though Miller recommended pieces of pots) to the bottom of containers to ensure that water drains through quickly. This advice is wrong.

Soil scientists determined long ago that water does not travel easily between soils with different textures, such as fine and coarse. When a finer-textured potting medium is situated above a thick layer of gravel, the finer material must be saturated before water will move into the gravel layer. Therefore, the potting medium will have to get soaking wet before water will drain into the gravel. A soggy medium is exactly what you do not want in a container.

WHAT HAPPENS IF YOU ADD ROCKS TO IMPROVE DRAINAGE

Putting gravel at the bottom of containers to improve drainage sounds like a good idea, but it actually makes the drainage worse. You can use gravel at the very base of a container to prevent soil from flowing out the holes or to add a little weight to the container, but that's about it. Anything more than a single layer of gravel will begin to affect drainage.

A BETTER WAY

If you do want to improve drainage, consider mixing perlite, a highly porous volcanic glass, into the existing container medium. Vermiculite, a form of expanded mica, will work as well, but it will not last as long.

━▶ the real dirt *Do not use gravel in the base of containers for anything but reducing the size of the drainage holes, or perhaps for adding a little weight.*

....

3

PEST, DISEASE, and WEED CONTROL

In This Chapter

Good Advice

∘ Use insecticidal soaps to control garden pests

∘ Apply pesticides in the morning

Advice That's Debatable

∘ Use homemade deer repellents rather than commercial ones

∘ Choose organic rather than synthetic methods of pest control

∘ Use homemade insect sprays rather than commercial ones

∘ Use corn gluten meal as a natural weed killer

∘ Release ladybeetles into your garden to help control pests

∘ Avoid using glyphosate because it lingers in the ground long after it has been sprayed

∘ Don't bother trying to control rabbits, voles, moles, and other small mammals

Advice That's Just Wrong

∘ It's okay to use outdated pesticides

There are all kinds of ways to get rid of pests around your garden and home. To begin with, though, we should tell you the method we prefer for controlling pests: nada, nothing, zilch. That's right, whenever possible, we prefer to ignore these problems. So what if insects eat a few leaves? Let a slug have a snack. Is it really a big deal if a disease causes a few spots? We just don't see what's so bad about a little bit of plant damage.

If the problem gets too big to ignore, before reaching for chemicals, we use our hands. Pull the offending weed, pick off the insect, pluck away those diseased leaves and throw them away. Only as a last resort do we consider other options. Some of these options are simple, logical things like planting a disease-resistant crop or using companion planting to deter pests or disease. More extreme measures might include using a homemade concoction or an organic or synthetic pesticide.

As you read this section, be aware that many of the chemicals we discuss are dangerous and, despite what you may have heard, could be potentially harmful to you, wildlife, and the environment. So take the warning labels on the pesticides you buy very seriously, and be cautious when using any homemade spray. All it takes is a little hot pepper in your eyes to absolutely ruin your day!

good advice Use insecticidal soaps to control garden pests

What is the safest thing to spray for pests in the garden? Again, we'd advise spraying nothing. If you are not up for that, try spraying water to knock pests off the plants. This works wonders with many pests, particularly aphids. No luck? Okay, try insecticidal soap, which works by washing the outer cuticle off of insects, causing them to dry out and die.

WHAT HAPPENS IF YOU USE INSECTICIDAL SOAPS

Sprays of insecticidal soap are safer than just about any other insecticide available, and they are effective against many soft-bodied pests such as aphids and mites.

If you decide to forgo using insecticidal soap to treat an insect problem, that problem will probably get worse, especially if the pests you are trying to control are on indoor plants where predators are unlikely to show up. Conversely, if you decide to use a more toxic pesticide, you are exposing yourself to potentially dangerous chemicals, while in all likelihood killing beneficial insects that could help the plants.

HOW TO DO IT

The easiest way to come by insecticidal soap spray is to buy insecticidal soap and use it as indicated on the label. If you'd like to make your own, you certainly can. We provide recipes for homemade insect sprays on page 78.

—➤ the real dirt *Insecticidal soap is a great cure for infestations of soft-bodied insects.* Too often gardeners jump the gun and reach for more toxic alternatives when a little insecticidal soap would do just fine.

good advice Apply pesticides in the morning

If you are going to apply pesticides, morning is widely considered to be the best time to do it. Winds tend to be calmer in the morning, so pesticides sprayed at this time are less likely to drift. Certain pesticides, such as sulfur, cause burns on plants at higher temperatures. Another benefit of morning application is that temperatures tend to be lower earlier in the day.

WHAT HAPPENS IF YOU APPLY PESTICIDES IN THE MORNING

To minimize the risk of inadvertently getting pesticides where they do not belong, as well as the risk of damaging plants, morning is the best time to apply pesticides—if you are going to apply them at all. Applying pesticides at other times of the day is not really a problem, except that it will be somewhat more difficult to find a time without wind, and you will need to be more careful about how hot it is.

HOW TO DO IT

In general, the hours between 6:00 and 10:00 a.m. are the best times to apply pesticide. You be the judge, though. If it's windy, don't spray. Carefully follow the directions on the label to ensure your safety and the safety of other living creatures, as well as that of the environment.

━● the real dirt *If you are going to apply pesticides, do it in the morning when temperatures are lower and winds have not yet kicked up.*

advice that's debatable Use homemade deer repellents rather than commercial ones

The commercial deer repellants you buy from a garden center usually include just about the same ingredients you could pull from a grocery shelf, such as hot peppers and eggs. Even some of the more exotic repellants, like wolf or lion urine, are not that difficult to simulate using your own urine or urine from a dog or cat. The question is, then, whether a homemade repellant is the equivalent of the fancier store-bought kind. The answer? It could be.

The smells and tastes that can come out of your kitchen can repel deer. Of course, what repels one deer might not be as effective on another. So it's hard to say whether a deer would dislike something you made or something you bought more.

Commercial repellants do offer one clear advantage over homemade: they tend to last longer after application. Manufacturers have accomplished this by using various compounds to make the bad-smelling stuff stick to the spot where it is applied rather than washing away, like homemade repellants do. Homemade repellants usually last for a week or two, whereas commercial repellants may last for anywhere from one to eight weeks. Of all the repellants available, egg is the ingredient that seems to be most effective. Still, because all deer are different, you will probably need to try several things to find the one that works for you.

WHAT HAPPENS IF YOU USE HOMEMADE DEER REPELLENTS

Homemade deer repellants are worth trying because they are inexpensive, easy to make, and there's a good chance they will work as well as commercial repellants, though they may not last quite as long.

A BETTER WAY

To make deer repellant, simply mix four eggs with a quart of water and add 2 or 3 ounces of a very hot pepper sauce (such as Tabasco). Blend it all together, strain, and spray it onto the plants you want to protect.

This concoction may burn foliage, so try it on a small area of leaves and wait a few days to see what effect it has on the plant before spraying it all over everything. In the meantime it can be stored in the refrigerator. But make sure to label it so it isn't mistaken for something edible! The odor from the application should spread, so that perfect coverage isn't necessary. Because these peppers are so hot, be careful to avoid contact with them, too.

Alternatively, if you do not want to deal with all this mixing, you could spread some urine around the base of plants and see how that works. Do not get urine on the leaves, though, because it will probably cause them to burn.

In addition to chemical repellants, there are also physical repellants for deer that you can buy. The most notable of these is the Scarecrow. When a deer walks by, the Scarecrow's motion detector, which is hooked up to a hose, ensures the intruder gets a startling spray of water. Fun for kids, scary for deer. What a great combination!

—▶ the real dirt *Both commercial and homemade remedies can successfully repel deer.* So it's really just a matter of how much you want to fuss around in the kitchen versus how much you want to spend.

advice that's debatable Choose organic rather than synthetic methods of pest control

If we were considering only organic methods of pest control that do not include chemicals, we'd put this one in the good advice section. But there are plenty of organic pesticides that include natural chemicals that could be dangerous. We fully support organic strategies for controlling pests, such as using beneficial insects or plucking bugs from plants by hand. What makes this advice debatable is that there are a whole bunch of organic poisons that are quite dangerous. Take copper sulfate and pyrethrum, for example. Both can be poisonous to us and the environment. Copper sulfate is very toxic to aquatic organisms, and pyrethrum is a broad-spectrum insecticide that can hurt beneficial and harmful insects alike. So we have to be careful about giving organic methods of pest control a blanket thumbs up. All too often organic is equated with safe, and this is not always the case.

WHAT HAPPENS IF YOU USE ORGANIC PEST CONTROL METHODS

There are plenty of good reasons for considering organic methods of pest control, particularly those that do not involve using chemicals. The majority of organic pest control methods are environmentally safe and less toxic than synthetic alternatives, except in the case of organic pesticides. Depending on which ones you choose, organic pesticides can be dangerous, and many gardeners do not realize it.

Don't want to use organic controls? Well, you could opt to do nothing about a pest problem. Environmentally this is appealing because you will not be spraying any potentially harmful pesticides. Unfortunately, if you do nothing the problem may reach a level where you'll have unacceptable losses, for example, no tomatoes.

You could also opt for a synthetic pesticide. These are relatively safe for humans and animals if used as directed. But, as with many organic pesticides, they often kill much more than the pest being targeted, including beneficial insects like bees and ladybeetles. Losing these pollinators and natural pest fighters (ladybeetles are great at getting rid of aphids) may actually lead to a pest problem that is worse than the original one, not to mention possible negative effects on the local ecosystem.

A BETTER WAY

If you are interested in controlling a particular pest, the first priority is to identify your enemy. Without proper identification, your efforts to control the pest are likely to be useless and you will be spraying chemicals without an understanding of their effects, potentially making the problem worse. Once you identify the pest, you can narrow down the methods of control to use.

Always start by looking at organic solutions that do not involve the use of pesticides. There are too many types of pesticide-free pest controls to mention here, but some of the best strategies include placing bags around young fruit, such as apples, to prevent insects and disease from infesting them as they mature; using traps of various sorts to catch insects (such as the Ladd trap to catch apple maggots and yellow sticky cards to catch aphids); and planting disease- and insect-resistant plants.

If these or other pesticide-free pest controls do not seem appropriate for your situation and you are not comfortable with the idea of taking no action at all, select the best pesticide you can whether it be organic or synthetic. Your goal should be to choose the pesticide that will do the least amount of damage to your health and the environment while still controlling the problem.

Check pesticide labels for information about the level of danger they may pose. A product marked danger is more toxic than one displaying the word warning, and caution is even less worrisome. The label will also tell you whether the pesticide you have selected can be used near water or wildlife. For more in-depth information about pesticides, both organic and synthetic, as well as pesticide-free organic choices, check out Jeff's previous book *The Truth About Organic Gardening: Benefits, Drawbacks, and the Bottom Line.*

━● **the real dirt** *There are a lot of wonderful organic techniques for controlling pests that do not involve the use of pesticides.* Use them! If these don't work for some reason, try to avoid having a predisposition toward selecting organic pesticides. They get touted as safe, but some of them can be just as dangerous as synthetic pesticides.

....

advice that's debatable Use homemade insect sprays rather than commercial ones

There are a lot of recipes for homemade insect sprays for controlling insect problems around your home or garden. These sprays are made of ingredients including soap, hot peppers, and garlic. Most of these homemade remedies are meant to repel pests rather than kill them, so they do not reach the same level of efficacy that commercial sprays do.

Of those intended to kill insects, soap sprays are often the most effective. They are also the safest for people and usually for plants. Sprays that include hot peppers can be effective, but they can also be very dangerous. The capsaicin that makes peppers hot can burn your eyes or even your skin if it's concentrated enough.

WHAT HAPPENS IF YOU USE HOMEMADE INSECT SPRAYS

Homemade insect sprays are usually convenient, easy to make, relatively inexpensive, and, at least for us, fun to try out. For example, we have tried homemade soap sprays on aphids and hot pepper sprays on mites and found them effective, though not as effective or fast acting as commercial sprays of synthetic poisons.

Forgoing homemade sprays in favor of more conventional ones probably will result in more dead pests. Forgoing any sprays may be the best choice of all. Though skipping sprays will often, in the short term, lead to more pests, in the long term it will allow predators to move in. This balance between predators and pests may still include more insect pests than you are comfortable with, but you will never know for sure unless you try.

A BETTER WAY

The better way to deal with insect pests may be to let nature take its course. However, if you want to try a homemade spray to get rid of insects, we recommend dish soap, hot pepper, and garlic sprays. Dish soap spray is effective for controlling soft-bodied insects, such as mites, aphids, and mealybugs. For more serious insect problems, such as beetles or caterpillars, hot pepper or garlic sprays may be more useful. Use caution, though. Hot pepper spray will cause a great deal of pain, and perhaps injury, if it comes into contact with your skin or eyes. Though not as dangerous, garlic spray also may cause injury if it gets into your eyes.

A simple dish soap spray can be made by adding 1 or 2 tablespoons of dish soap (not the detergent you'd use in a dishwasher) to 1 gallon of water. Before using this spray, apply a little bit to a small area of a plant and let it sit for a couple of days to see whether it causes any damage. If all is well, the whole plant can be lightly sprayed.

To make a hot pepper spray, mix 1 cup of the hottest peppers you can find, such as habaneros, with ½ cup of water in a blender. Strain out any solid parts, and mix the liquid that's left over with a little more water and 1 tablespoon of dish soap. (Of course, the more water you use the less concentrated the spray will be.) We recommend diluting the mixture with about 2 cups of water to start. Again, test the spray on a small area and then observe whether any damage appears after a couple of days before using it on a more widespread area.

To make a garlic spray, blend 1 cup of chopped garlic with ½ cup of water. Strain out any solid bits, and then add 1 tablespoon of dish soap. Adjust the concentration of this spray by diluting it with water (again, we recommend starting with about 2 cups of water). Test it on a small part of the plant before use.

None of these homemade sprays will last very long after spraying, so you may need a second application four to seven days afterward to clean up the remaining critters.

━● the real dirt *Using homemade sprays can be fun and economical.* It's important, though, to be careful not to injure yourself or others. You also need to be okay with injuring a plant or two during your tests. Do not expect the kind of instant gratification you get with commercial sprays.

advice that's debatable Use corn gluten meal as a natural weed killer

In the world of organic herbicides, the most sought-after product is one that will effectively control weeds before they emerge.

So far, the only product to come close to doing this successfully is corn gluten meal. A by-product of the corn industry, corn gluten meal is touted as not only helping to stop weeds, but also as a way to provide some fertilizer to plants. (It is a 10–0–0 fertilizer, which means it contains 10 percent nitrogen and no phosphorus or potassium.)

Corn gluten meal controls weeds in two ways. Most notably, it contains certain chemicals that prevent seeds from germinating properly. Because of its nitrogen content, corn gluten meal can also provide a boost for plants already growing in a location. Though it can be used on gardens, it is usually recommended for lawns. In a lawn it will help fertilize the grass to the point that a weed seed cannot find a free spot to germinate.

The biggest drawback to this product is that it usually takes two years for it to be effective and even then, it will only control about 80 percent of your weed problems. This is far below what you can expect from synthetic herbicides.

WHAT HAPPENS IF YOU USE CORN GLUTEN MEAL

Corn gluten meal is a relatively safe thing to add to your lawn or garden. It has very low toxicity to humans and is not known to harm anything in the environment beyond weeds, although it is possible to apply too much and pollute nearby bodies of water. Corn gluten meal not only helps to control weeds by keeping weed seeds from germinating, it also provides enough nitrogen (when applied at the recommended rate) for your lawn or garden for an entire year of growing. Be aware, though, that corn gluten meal does not work on existing weeds. It only works on seeds.

Obviously, this product is not a good choice if you are the type of gardener that likes to let perennials go to seed because it would also stop these seeds from germinating. If you decide to forego corn gluten meal, you will need to control weeds in some other way. We recommend hand weeding whenever possible, but others prefer to use synthetic chemicals.

A BETTER WAY

For controlling weeds in a garden or lawn, hand weeding is almost always the best choice if you have the time and energy. If you don't, corn gluten meal can provide some help. Corn gluten meal products sold for controlling weeds include instructions on how to apply them. An application of 20 pounds per 1000 square feet is commonly recommended. This may be applied at one time, but more frequently the application is split in half for spring and autumn delivery. Immediately after applying it, water the corn gluten meal in lightly. If you use corn gluten meal in the garden to control weeds, make sure the seedlings you are growing have had enough time to establish themselves before applying it.

━▶ the real dirt *Corn gluten meal is a safe, effective, pre-emergent herbicide to control weeds, but it is not as effective as synthetic chemicals.* Though it takes some time to do its job, we recommend trying corn gluten meal and hope more people will look into using it.

advice that's debatable
Release ladybeetles into your garden to help control pests

When we talk about beneficial insects, most people first think of ladybeetles (better known as ladybugs). There's a good reason for that: ladybeetles are excellent predators. They feed on insects that eat our vegetables and ornamental plants, such as aphids and mites.

The problem with releasing ladybeetles in a small yard or garden is that they like to fly when they are let go. Even if there is plenty of food available, ladybugs are predisposed to spread themselves out rather than congregating in a single spot. So there's a good chance that most of them will be gone shortly after you turn them loose.

WHAT HAPPENS IF YOU RELEASE LADYBEETLES
Ladybeetles are nice to have around. They are great predators, and even if they do not stay in your garden it's not as though you have released a plague into your neighborhood. In fact, there's a good chance you have helped out the neighbors more than yourself.

A BETTER WAY

If you decide not to release ladybeetles but still want to try to control pests naturally, there are other predatory insects you can try. Two of our favorites are the minute pirate bug *(Orius* species*)* and the big-eyed bug *(Geocoris* species*)*. Green lacewing larvae *(Chrysoperla rufilabris)* are another excellent choice.

Releasing beneficial insects, including ladybeetles, is simple. Just buy the insects you want from a reputable source—usually online, as few garden centers stock them for logistical reasons. Once you receive them in the mail, release predators near the insects you would like them to feed on.

—● the real dirt *We are not huge fans of releasing ladybeetles because, in our experience and the experience of most researchers and gardeners we know, they never stick around.* However, some people have had excellent success releasing ladybeetles into greenhouses or conservatories.

advice that's debatable Avoid using glyphosate because it lingers in the ground long after it has been sprayed

The most common garden weed killer in the United States is glyphosate, which is the active ingredient of the herbicide Roundup, as well as many other popular brands. When glyphosate is sprayed onto a plant's leaves, the chemical is translocated by the plant to its root system, where it prevents the plant from producing an amino acid necessary for survival. This chemical has been around for more than 40 years and has been studied widely for its safety and longevity in the environment. One of the major criticisms of glyphosate is that it can remain in the soil for a long time, potentially years.

It's true that glyphosate does bond very quickly to soil particles. But once it bonds to these particles it is deactivated. Because bonding occurs rapidly, it is generally safe to plant in an area where glyphosate was applied just a few days after the application. However, the bonded glyphosate could remain in the soil—albeit in an inactivated form—for years.

There is no evidence that glyphosate that is bound to soil particles has any harmful effects on the environment. If misused, though, through drift or inadvertent application to areas where it should not be applied, such as streams or other waterways, it can adversely affect a wide variety of living things, including aquatic insects, tadpoles, and, of course, plants that are unintentionally treated.

WHAT HAPPENS IF YOU USE GLYPHOSATE

There are plenty of good reasons to avoid using glyphosate, especially its potentially harmful effect on water-dwelling creatures if it is misapplied. But staying away from it because it lingers in the soil is not one of them. Glyphosate does linger in the soil, but because it is so tightly bound there it cannot actually do anything.

If, despite the pleading of others, you decide to apply glyphosate around your garden or yard, you need to be careful not to let it come into contact with plants you want to keep, waterways, people, or animals. Follow the directions on the label carefully.

A BETTER WAY

Rather than using glyphosate, use your hands and hoes to dig up offending plants.

━▶ the real dirt *We use glyphosate periodically, but it is not our first line of defense against weeds.* If you do use glyphosate, or any other herbicide, do so carefully and according to the instructions on the bottle. Do that, and there should be no significant side effects.

....

advice that's debatable Don't bother trying to control rabbits, voles, moles, and other small mammals

Small mammals are some of the toughest critters to control. Despite a flood of advice, there is not a sure-fire cure that keeps small mammals out of your garden. Still, there are some things that you can try. A homemade deer repellant made of eggs, water, and hot pepper sauce (see the section on deer repellants) could help. So might some of the commercial repellants. Ideally it's nice to think that you can share your bountiful garden with rabbits, but in practice this rarely ends up working out.

WHAT HAPPENS IF YOU USE REPELLANTS AGAINST SMALL MAMMALS

Small mammals can and will feed on tree bark and roots of perennials and trees, and they may even steal your vegetables. Repellants are worth a try because they can work and they do not kill the pests (assuming you have an aversion to killing mammals). If you have a serious problem with mammals, you're going to need to do something about them or you could lose most of your vegetables in the summer and, potentially, young trees in the winter. If repellants are not working, the alternatives are nasty poisons, fencing, or live traps (though placing a small mammal somewhere far from its home in the dead of winter doesn't usually result in that animal remaining alive for long). Fencing often works for a while, but the mammals usually find a way around it eventually. And poisons are toxic to just about everything. Sure, poisons are usually locked in some kind of structure that a dog, cat, or child should not be able to get to, but these structures are not foolproof.

A BETTER WAY

Try everything! Just because one repellant doesn't work, that does not mean they all won't. Try the homemade deer repellents we mentioned on page 73. Also try fertilizers like milorganite and blood meal, which have been known to repel small mammals. Experiment with predator urine and commercial compounds. The key to finding something that works is having the patience for trial and error.

If you do go with a commercial repellant, follow the instructions on the label carefully. These compounds can be dangerous, and care needs to be taken. Castor oil is an ingredient in commercial repellants that has produced good results.

—● **the real dirt** *Try to be tolerant if small mammals are causing a little bit of damage in your garden.* If the problem becomes serious, though, there are ways of preventing the damage. First, try various repellants or live traps. Use poisons only as a last resort and with strict attention to instructions on the label.

advice that's just wrong It's okay to use outdated pesticides

Pesticides, just like almost any other chemical, go through changes over time. Heat, cold, humidity, exposure to air, and the properties of the pesticide itself all affect how long a product lasts. Fortunately, some pesticides have use-by dates on them. We advise paying close attention to these dates to avoid using a pesticide that has changed chemically.

If there is no use-by date, it is safest to assume that a pesticide will only last for about two years after purchase—assuming that you do not allow the pesticide to get very hot or very cold, because allowing pesticides to heat or freeze can ruin them. If you want to use the pesticide after two years, you should contact the company that produced it and find out what they say about the life of the pesticide.

WHAT HAPPENS IF YOU USE OUTDATED PESTICIDES

Do not consider using old pesticides. Not only will they almost certainly be less effective than they should be, they may also be more toxic. In some cases, they may no longer even be legal as many older pesticides are now banned.

A BETTER WAY

If you own a pesticide that is over two years old, contact the manufacturer to find out how to dispose of it. Many counties have hazardous waste disposal units that also can be contacted. Because pesticides only last so long, when purchasing pesticides you should only buy the amount that you need—despite what a good bargain a larger quantity might be.

—• the real dirt *Jeff once knew a man who applied DDT 30 years after it had been banned.* While there was never a lawsuit filed or any known damage, this was obviously not a good idea for lots of reasons. Do not use old pesticides, particularly those that have been banned.

4
MULCH

In This Chapter

Good Advice

○ Use organic mulches rather than inorganic mulches for most garden beds

○ Keep mulch away from the crowns of ornamental plants and the bases of trees

Advice That's Debatable

○ Always mulch gardens

○ In cold climates, apply winter mulch to perennials after the ground has frozen

○ Use landscape fabric to control weeds

Advice That's Just Wrong

○ Avoid wood mulch because it attracts carpenter ants and termites

○ Always add extra nitrogen to the soil when wood mulch is used

Mulch is a layer of material that is spread over the soil surface. When used properly, mulch not only conserves moisture and suppresses weeds, it also helps protect plant roots from the elements and keeps soil from eroding or blowing away. As it breaks down, mulch releases nutrients that enrich the soil. It can also reduce the amount of water and soil that splashes onto lower leaves during a rainstorm, which can lead to disease problems.

If you want to minimize the amount of weeding, watering, and raking you have to do, your best bet is to cover garden beds with mulch. Sounds easy, right? Well, sort of. There are lots of different types of mulch, so some care needs to be taken when choosing what's best for the conditions in your garden. Choices range from organic mulches, such as straw and wood chips, to inorganic options, such as gravel, plastic sheeting, and recycled rubber. How you apply mulch often depends on the time of year and what you have planted. This is where most of the debate and confusion comes in.

good advice Use organic mulches rather than inorganic mulches for most garden beds

To be honest, this could have gone in the debatable section because there are so many differing views on this topic. In our experience, though, using mulches made from plant sources are the best choice for garden beds because they serve multiple purposes.

Yes, organic mulches, such as wood chips, compost, bark, grass clippings, and straw (not hay, which can contain unwanted seeds) help keep weeds at bay and do all the other great things we mentioned above. But they also act as a slow-release fertilizer, adding organic material and nutrients to the soil as they break down. This not only improves the soil, it boosts plant growth, too.

Inorganic mulches, such as rocks, gravel, and plastic sheeting, have many disadvantages. First, in our opinion, they're ugly. They are hard to weed and keep free of debris. Inorganic mulches retain heat in the hot sun, often cooking plants' crowns, and light-colored stones can reflect sun and scorch leaves. And just try getting rid of them once you realize how dreadful they are. One positive, though, is that rocks and gravel are good choices for paths.

WHAT HAPPENS IF YOU DON'T USE ORGANIC MULCH

We know our share of gardeners who do not use mulch, organic or otherwise, for aesthetic reasons. "Bark and wood chips are ugly," they say. There is some truth to that, but if you don't use mulch you will need to do a lot more work to maintain a healthy garden with good soil. Unmulched soil dries out fast. In hot weather it can crack and harden in ways that resemble a desert rather than a place where living things thrive.

Your garden will not fall to ruin if you do not use organic mulch. But you will have to work harder to keep soil and plants healthy. If you are willing to do a lot of extra weeding and watering, you will probably be fine as long as you work compost or other types of organic matter into the soil on a fairly regular basis. This is a tradeoff made by many gardeners who do not like the look of mulch. So it comes down to aesthetics versus back pain and higher water bills.

HOW TO DO IT

Before you mulch, pull out the weeds. Do not try to get rid of existing weeds by covering them with organic mulch. They will just grow right up through it.

If bags of mulch are not in your budget, contact your city's forestry or parks department to see whether arborists' tree trimmings or composted yard debris are available for free in your area.

When to apply the mulch depends on your goal. If you are trying to keep soil cool and moist in the summer, apply new mulch or add mulch to existing mulched beds in the early spring. If you are growing vegetables and want to mulch the beds, wait until the soil has warmed up a bit before applying it. Vegetables like soil that is around 65°F (18°C), and if you apply mulch before the soil is close to that temperature you will be keeping the ground too cool for good growth. If you want to help insulate plants for the winter, apply additional mulch after the first frost once the ground has begun to freeze.

Opinions on how deep mulch should be vary from 2 to 4 inches. More than 4 inches is probably overdoing it for most situations. If you are applying mulch to trees, mulched areas should not be mounded up around the tree's trunk volcano style, as you so often see around newly planted trees. Keep mulch at least 3 inches away from plant stems and tree trunks.

—▶ the real dirt *Organic mulches replenish the soil in the same way nature would. (Think fallen leaves.)* Rocks, gravel, and recycled rubber products make great paths, but they are not the best choice for covering soil where plants need to thrive.

good advice Keep mulch away from the crowns of ornamental plants and the bases of trees

Because mulch has so many good characteristics, some garden-
ers think "the more the better." But this is not true. Mulch piled up
around tree trunks is a good example of this approach. Rather than making a deep cone when you mulch a tree, spread mulch in a circle around
the base of the tree to a depth of about 4 inches. Make the circle wide
enough to include the drip line of the tree (that is, the outline of the canopy), and keep the mulch away from the trunk. Mulching too deeply and
too close to trees is one of the quickest ways to injure or kill them. The
same is true of the crowns (where the stems meet the roots) of perennials
and other plants.

WHAT HAPPENS IF YOU DON'T KEEP MULCH AWAY FROM THE BASES OF PLANTS

When mulch is mounded deeply against a plant's trunk or crown, a number of bad things can happen. Roots can suffocate without the oxygen they
need. Too much mulch can also keep water from penetrating the soil the
way it should.

Mulch traps moisture when piled against a tree, perennial, or other
type of plant, creating the perfect breeding ground for fungal diseases.
Thick piles of mulch are also problematic because they can be an in-
vitation for rodents to take up residence there and start chewing on
bark and girdling trees.

HOW TO DO IT

Mulching around tree trunks not only prevents weeds from competing
with tree roots, it also goes a long way toward protecting them from string
trimmer and mower damage. Keep mulch a minimum of 3 inches from
the base of young trees and 6 inches or more from mature trees. For best
results, extend the mulch all the way out to the tree's drip line, or canopy

edge. (We know this will not work in many home landscapes, but do what you can.) When mulching ornamentals, apply it in a ring around the base.

━● the real dirt *To avoid suffocating plant roots, inhibiting moisture and air flow into and out of the soil, and fostering fungal diseases, mulch should be applied to a depth of no more than 4 inches and kept away from the crowns of perennials and the bases of shrubs and trees.*

advice that's debatable
Always mulch gardens

Despite all its benefits, there are some who just don't like the look of mulch and will not use it, preferring to work hard to pull weeds and enrich the soil with compost and fertilizers. That is a fine choice. But aesthetics aside, there are other times when mulch can and maybe even should be avoided. For instance, mulch provides cover and a breeding ground for insect pests and snails. Wet mulch spread too close to plants and trees can cause rotting, and mulch used on poorly drained soil can make the situation worse. If you mulch too early in the spring, the soil will likely warm more slowly than if it were left bare, slowing the growth of vegetables. For those who find weeding cathartic, mulching will significantly cut down on your time spent on this restorative activity.

WHAT HAPPENS IF YOU MULCH GARDENS

Mulching your garden helps to suppress weeds, conserve moisture, keep soil from eroding, and prevent water splash that can foster disease. Over time, mulch releases nutrients and organic matter that enrich the soil.

One of the best reasons for not using mulch is reseeding. Plants can reseed through mulch, but they don't do it as easily as on bare soil. Gardeners who want a naturalized garden may want to go without mulch, or rake it out of the way when plants go to seed and then cover the soil once seeds have dropped. Mulch should also be avoided in gardens where you are trying to encourage groundcovers to spread. By not mulching, you'll have more work to do in the garden to keep it healthy, but you may like the look better and you will likely see a lot more seedlings popping up than you would if you mulched.

A BETTER WAY

It's important to determine right from the start whether you like the look of mulch or not and, if not, if you are ready to deal with the labor that goes along with avoiding it. If you do not mulch, be mindful of the need to weed regularly. Plants and weeds compete for water, nutrients, air, and light. If weeds take over, your plants will suffer. Unmulched soil will retain moisture less efficiently, so you will need to water more often when it does not rain. Because mulch adds nutrients to the soil as it breaks down, without mulch you will need to add compost and other organic material to the soil from time to time to keep it healthy.

—● the real dirt *Mulch offers many benefits and is a good idea in most situations, but it is not a must for every garden.*

advice that's debatable In cold climates, apply winter mulch to perennials after the ground has frozen

We put this advice in the debatable section because, while it is useful, many experienced gardens do not follow it and do not suffer much for skipping it, Meleah included. The reason is simple: applying winter mulch can be a pain. Sure, if you do not apply it you may lose some tender perennials, but that's just survival of the fittest.

To practice winter mulching, remove any diseased plants and/or foliage, wait until the ground freezes, and then apply 2 to 4 inches of pine needles, shredded leaves (whole leaves can mat down and cause rot), or straw to the ground beneath perennials, shrubs, and young trees. While many think that the purpose of winter mulching is to keep the ground warm, it is actually used to keep the ground evenly frozen until spring.

WHAT HAPPENS IF YOU APPLY WINTER MULCH

Winter mulch serves as an insulator that keeps soil temperatures consistent and is most useful with herbaceous perennials that are considered tender and at the edge of their ability to tolerate the cold. When applied correctly, winter mulch ensures plants do not experience the full effects of extremely cold air temperatures, particularly when temperatures drop before there is a good blanket of snow covering the ground. Winter mulch also helps ensure that plants don't emerge too early in the spring from a brief warm spell, as it will keep the ground cooler than the surrounding bare soil.

This insulating layer of mulch also helps protect plants from frost heaving, which can happen when the soil freezes and thaws between late autumn and spring. As the soil shifts in response to these freeze–thaw cycles, plants can literally be heaved out of the ground, some to the extent that their roots are exposed and damaged. Winter mulch is also said to help plants retain at least some moisture during the coldest, driest months.

If you do not apply winter mulch, you may lose some tender or marginally hardy plants or plants you put in the ground late in the season. We were hard-pressed, though, to find even a handful of Minnesota gardeners who had ever had a plant actually heave—and it's really cold here. So it's unlikely that an entire garden will be lost if you do not use winter mulch. You can get by without applying winter mulch and lose few, if any, plants—even the more tender ones—if you do not cut down perennials you're trying to protect in autumn. As tender shoots and leaves from these plants die, this growth will fall around the plant's base and help protect it.

A BETTER WAY

Gardeners are always told to hold off on applying winter mulch until the ground has frozen. But this advice is not so cut and dried, really. If you are applying mulch to an area that does not drain well, you definitely want to wait until the ground has frozen. That way you will not trap moisture that could lead to disease.

While you do not want to apply winter mulch too early because it can cause the soil to retain heat and delay the freezing process, it is okay to apply once temperatures have gotten cold but the ground is not quite frozen yet. Even waiting until there's been just one hard frost is probably fine.

In some cases, though, it is be best to apply winter mulch before the ground has started to freeze. Some young trees and even a few shrubs known for having difficulty making it through their first winter (magnolias come to mind) benefit from early winter mulching. This insulating blanket of mulch helps protect their root systems, which are trying to get established, from being injured or even killed by deep-penetrating frost.

Once you have applied your mulch of choice to garden beds, pull it about 3 inches away from the stems and trunks of plants to decrease the risk of disease. This will also help deter critters that might burrow into the mulch from chewing on the bark and stems.

—▶ the real dirt *Mulching for winter is a good idea, but unless you have marginally hardy or recently planted perennials, shrubs, or trees, it probably won't cause a lot of problems to skip it.* You just have to decide whether you are okay with a may-the-strongest-survive attitude when it comes to your garden, or if you are willing to do a bit more babying for a few of your plants.

advice that's debatable Use landscape fabric to control weeds

Initially sold as a product to be used behind boulder walls to control soil erosion, fabric weed barriers are now touted as being a great way to eliminate weeds in the garden. But research has shown that fabric barriers labeled as "porous" can be harmful to plants and soil because they may not allow water to penetrate well enough. These barriers may also deprive soil and plant roots of the oxygen they need, and they are really difficult to remove when you'd like to move or add plants, trees, or shrubs. The biggest problem, however, is that these fabric barriers will get worn and damaged, and weeds will eventually sprout up through them anyway. When this happens it will be tougher to pull them out by their roots because the barrier will hold the roots tighter than the soil would. Yes, these fabrics do control weeds for a few years, but the eventual cost of these barriers in terms of time and stunted plants usually makes them more trouble that they are worth.

WHAT HAPPENS IF YOU USE WEED BARRIERS

Some gardeners use weed barriers to cut corners. And, really, who wouldn't love to eliminate the drudgery of weeding if they could? But these barriers just don't work forever. The worst weed barrier by far is the one you see most often: black plastic sheeting. We have even seen this material used on hills where, in a heavy rain, it literally becomes a slippery slide for mulch and everything else that is perched on it.

The worst thing about these barriers is that they often let very little, if any, water reach the soil, save for areas around the holes cut for plants. Without air or moisture, microbes that keep the soil healthy die, and plants that do survive in this environment certainly do not thrive.

By not using these barriers, you will have to weed, but you will have a much happier, healthier garden. Microbes will thank you and earthworms will sing your praises. If you want to decrease the amount of weeding you have to do, use mulch. Even inorganic mulch is generally better than these fabrics—but don't use them either because there is no hell like the hell of weeding around white rocks.

A BETTER WAY

Organic mulches are a good alternative to landscape fabric. If you are going to use a landscape fabric despite our warnings, however, select one that will allow moisture through it. For aesthetic reasons, it is usually best to cover landscape fabric with something a little more attractive. Either organic or inorganic mulches will work, though organic mulches will not be able to feed the soil as they normally would because the landscape fabric will block the decomposing organic matter from reaching the soil below it. When using landscape fabric, you will only need about 2 inches of mulch because the fabric will do most of the work of controlling the weeds—that is, until the weed barrier breaks up and decays.

—• the real dirt *Weed barriers can control weeds for a relatively short period of time, but we do not recommend using them.* They are a shortcut you will likely regret. Instead, spread a layer of organic mulch directly over the soil in your gardens.

advice that's just wrong Avoid wood mulch because it attracts carpenter ants and termites

Carpenter ants are found around homes, and they do bore through wood, such as living or dead trees or logs, to build nests. They do not eat wood, though, so they are not especially drawn to the wood chips and shredded bark we use as mulch in our gardens. Numerous studies have confirmed these facts, and yet this myth continues to circulate over backyard fences, as well as the Internet.

It is true that termites are fond of wood and are often found in it. But entomologists at universities around the country say that termites are not attracted to wood mulch. In favorable climates, the workers of the termite community tunnel to the soil surface to feed on wood and carry it back to the colony for others to enjoy. Moist environments create more favorable conditions for tunneling, so in that sense the presence of wood mulch (or any kind of mulch) can be favorable to termites. But researchers at the Structural Integrated Pest Management Program at the University of Maryland found that termites do not feed much on the types of hardwood found in mulch. According to Donald Lewis, an entomologist at the University of Iowa Extension Service, termites were found just as often beneath mulches of hardwood, pine bark, eucalyptus, and pea gravel as they were beneath uncovered, bare soil.

Some of you may have heard rumors about Formosan termites being in mulch made from trees felled by Katrina and other hurricanes in New Orleans. These rumors are not true either. As Lori Bushway and Carolyn Class, senior extension associates with Cornell's Cooperative Extension, put it, "Termites do not have a very long half-life in a shredder, so fresh mulch is not a problem." As for bagged mulch that sits around a while, they said it is possible that termites could get into those bags lying directly on the ground. "But that has been going on for decades and involves local termites."

WHAT HAPPENS IF YOU USE WOOD MULCH

If you want to use wood mulch but have been afraid to do so because of fears of attracting carpenter ants and termites, you can let those fears go and mulch to your heart's content. Wood mulch provides organic matter for the soil, helps keep the soil moist when the weather is dry, and stops weeds from sprouting. In other words, it's an asset not a hazard.

A BETTER WAY

Entomologists do have some cautionary advice for gardeners regarding carpenter ants and termites. It's always a good idea to keep wood mulch several inches away from the foundation or your house and other buildings. Mulch should also not come in contact with siding. If you do spot termite activity around your property, call an exterminator.

→ the real dirt *Do not be frightened by this myth.* Research does not support the connection between carpenter ants or termites and wood mulch.

....

advice that's just wrong Always add extra nitrogen to the soil when wood mulch is used

This debate just won't die: "Wood chips tie up nitrogen in the soil, so you must add nitrogen whenever wood mulch is used near plants," one camp argues. "No it doesn't," the other camp retorts. People think that nitrogen will disappear from soil if they use wood mulch because microbes use nitrogen to break down fresh wood.

The truth is that some nitrogen from the top layer of soil is used up when wood chips are initially broken down by bacteria. But this nitrogen loss is likely to affect only the most tender annuals, and even they probably will not sustain any permanent injury.

WHAT HAPPENS IF YOU ADD EXTRA NITROGEN

If you want to add nitrogen when mulching, you certainly can, but do not add any more than recommended on the fertilizer label. Even if you do not fertilize, it is wise to mulch because it provides so many benefits to soil and everything that lives in it.

The confusion over whether wood chip mulches cause nitrogen deficiencies that harm plants persists, despite numerous studies demonstrating that serious depletion of nitrogen is not taking place. In fact, many studies have indicated that just the opposite occurs. As wood mulches decompose, they enrich the soil with many nutrients, including nitrogen. Because this is the case, adding excess nitrogen when none is needed could do more harm than good. If you are concerned about the health of your plants and believe they may not be growing well due to a lack of nitrogen, get a soil test done.

A BETTER WAY

Wood mulch, either bagged or delivered fresh, can be used anywhere in the garden. Spread mulch to a depth of about 3 to 4 inches. If you will be adding fertilizer as part of your regular plant care ritual, it is best to apply it prior to mulching so the fertilizer can more easily infiltrate the soil.

━▶ the real dirt *Gardeners who use wood chips do not need to add additional nitrogen to the soil beyond what they normally add to ensure plant health.* While some nitrogen depletion does occur as the wood chips decompose, it is not significant enough to cause problems. Over time, wood mulch adds more nitrogen to the soil than it takes.

5

ANNUALS, PERENNIALS, and BULBS

In This Chapter

Good Advice

○ Deadhead to encourage bloom

○ Choose plants that are suitable for your hardiness zone

○ Harden off seedlings before transplanting them outdoors

○ Separate the roots and remove heavily matted clumps of root-bound plants before planting

○ Use non-natives as well as natives in a landscape

○ Plant perennial beds rather than turf grass beneath trees

○ Plant disease-resistant cultivars when you can

○ Plant perennials in spring and early autumn

Advice That's Debatable

○ Follow spacing recommendations on plant labels

○ Use expensive grow lights to start plants successfully from seed indoors

○ Do not place plants labeled "full sun" in spots that get less light

○ Wash and sterilize containers at the end of the season

○ Divide plants and transplant only in the spring and autumn

Advice That's Just Wrong

○ Add phosphorus to increase bloom and stimulate rooting

As gardeners, we are always searching for new ways to use annuals, perennials, and bulbs effectively in our gardens. Each type of plant offers its own set of advantages. Sure, we have to replant annuals every year, but their long-lasting blooms make that work worthwhile. Perennials, on the other hand, generally flower for just a short time during the season, but they last for years and eventually mature into what amounts to the skeleton of the garden. And there's just something magical about bulbs. They are easy to plant, do not need much care, and most return reliably year after year (with the exception of tulips) and look fantastic.

So what's so confusing about these three types of plants? Plenty. Read any garden magazine or tune into any radio or television show on annuals, perennials, and bulbs and you'll likely be told something different about how to choose and plant them, as well as how to care for them. Particularly with the plants covered in this chapter, some tips sound so reasonable, yet so often they are misguided or wrong.

Here's a good example: buy perennials in gallon containers or larger because they will be less fragile and easier to care for than smaller ones, making them more likely to do well in the garden. This sounds logical, doesn't it? The truth, though, is that younger, smaller plants are less apt to suffer transplant shock than larger plants. The lasting effects of transplant shock often causes larger, more mature plants to drop leaves or produce stunted leaves and stems. It's also common for them to grow really well for a while before seeming to just stop. If it's well cared for, in two years that perennial you bought in the 4-inch pot is likely to be thriving and just as large, if not larger, than the gallon-size version of the same plant you bought for more than twice the price. That's why it's better to buy small perennials.

If you must have large plants because you crave instant gratification or are hosting a party in your yard, you can minimize transplant shock by making sure plants are not root-bound before you buy them. Do this by gently sliding the plant out of the pot into your hand. (It's okay to do this, and if the place you are shopping objects, go somewhere else.) If the roots look almost white and they are not encircling the pot, you've got yourself a keeper. If they're brown, black, or smell bad, move along.

good advice Deadhead to encourage bloom

It's kind of a creepy term, but deadheading is actually a good thing and it means exactly what it sounds like. Once a plant's flowers are spent, meaning they are looking pretty scraggly but are not yet dry, snipping, pinching, or shearing off those dead flower heads will lead to more blooms. You need to deadhead most flowers more than once per season. Some flowers, such as *Astilbe,* bearded iris *(Iris germanica),* Siberian iris *(Iris sibirica), Hosta, Penstemon,* lamb's ears *(Stachys byzantina),* peonies *(Paeonia),* lungwort *(Pulmonaria),* goat's beard *(Aruncus dioicus),* and poppies *(Papaver),* will not rebloom whether you deadhead or not. Still, getting rid of worn-out flowers will make these plants look better in most cases. It's your preference, really. We like the dried heads of Siberian iris, poppy, and *Astilbe*, so we leave them alone. Deadheading can be helpful for both annuals and perennials because it encourages the plants to divert their energy from producing seeds to producing flowers.

WHAT HAPPENS IF YOU DEADHEAD

It takes a lot of energy for plants to produce seeds. By removing old flowers before they go to seed, the plant can focus all its energy on new growth.

If you do not deadhead, you won't get those extra few weeks or months of bloom time that you would have otherwise. By letting flowers go to seed, you will also get a lot more volunteers in the garden as plants self-seed wherever they like. If this is something you want, let plants go to seed and cut them back only when they start to look really bad. A few good self-seeders are columbine *(Aquilegia),* bachelor buttons *(Centaurea cyanus),* bugleweed *(Ajuga),* lady's mantle *(Alchemilla vulgaris),* Queen Anne's lace *(Ammi majus),* yarrow *(Achillea millefolium),* coneflower *(Echinacea purpurea),* and sweet viola *(Viola odorata),* to name just a few. You may also choose to leave dried seed heads to provide food for birds.

HOW TO DO IT

Deadheading methods differ, depending on the plant. Some explanations of the practice can really make your head spin because of all the jargon and complicated illustrations they include. Common sense is your best guide on this one. You will know it's time to deadhead when blooms are well past their peak. But it is best to deadhead before blooms get dry and the plants have already worked hard to set seeds.

Use scissors or shears to deadhead low-growing, bushy plants, such as hardy geranium *(Geranium sanguineum)*, pinks *(Dianthus)*, white and blue clips *(Campanula carpatica* 'White Clips' and 'Blue Clips'), and threadleaf coreopsis *(Coreopsis verticillata)*.

Spent flowers on long or woody stems should be snipped with a hand pruner just above a set of leaves or where another stem joins the plant. Other flowers can be easily removed individually either by pinching them between your fingernails or snipping them off with scissors or a pruner. Regardless of the way you do this, remove flowers at the base of a stem so you are not leaving headless stubs behind, which looks much worse than scraggly flowers.

A few plants that do well when pinched are pincushion flower *(Scabiosa columbaria)*, oxeye daisies *(Chrysanthemum leucanthemum)*, yarrow *(Achillea millefolium)*, larkspur *(Delphinium)*, daisies *(Chrysanthemum)*, cardinal flower *(Lobelia cardinalis)*, salvia *(Salvia splendens)*, speedwell *(Veronica officinalis)*, tickseed *(Coreopsis)*, balloon flower *(Platycodon grandiflorus)*, blanket flower *(Gaillardia aristata)*, and daylilies *(Hemerocallis)*.

━▶ **the real dirt** *Deadheading is a quick, easy way to get plants that rebloom to keep flowering for weeks or months.* But don't leave a bunch of headless stems behind.

good advice Choose plants that are suitable for your hardiness zone

If it matters to you whether the plants in your landscape survive (or, at least, are likely to survive) from season to season, you will want to choose the ones deemed suitable for your hardiness zone. These hardiness zone classifications are based on the average minimum temperature that a plant will experience over the course of a year. A map of the hardiness zones in the United States, Canada, and Mexico is available at the website of the U.S. National Arboretum (http://www.usna.usda.gov/Hardzone/ushzmap.html).

All plants are classified according to their hardiness zone, and you will find zone information on the plant's tag or container label. Plant hardiness zones are expressed in a range. For example, a California redwood tree is hardy in zones 5 to 8, whereas a coneflower grows reliably in zones 3 to 9.

WHAT HAPPENS IF YOU BUY PLANTS BASED ON HARDINESS ZONES

If you want the plants that you've invested time and money in to last, it's important to pay attention to the hardiness zone when you are shopping. This is most important when you are making major purchases, like trees.

Annuals are a good example of plants we choose without regard to hardiness zone. We buy annuals for our gardens knowing they will not survive the winter, but they are nice to have around while they last and many go to seed. It can also be fun to experiment with plants that are right on the edge of a zone to see how they will do. If you live in zone 4, say, and you are dying to have a Japanese maple, which is usually rated as a zone 5 plant, you might be able to pull it off if you plant it in a sheltered spot.

HOW TO DO IT

Always look for hardiness information when you are shopping for perennials, shrubs, and trees. Be aware that signage made by sellers, even at reputable places, may sometimes stretch the truth a bit. We have seen signs urging shoppers to buy mums labeled as hardy that were never meant to survive a Minnesota winter. Base your decisions on the actual nursery tag or plant label.

New introductions can also be problematic because, although they may be rated for a particular zone, they are new, so it's still up in the air how they will fare over time. You also have to consider that hardiness zone maps are far from perfect. For example, while the U.S. version does a good job of taking cold into consideration, it is often criticized for not accounting for the extreme heat in the West and South. In addition, the effect of cold on a plant is also influenced by other things, like snow cover or whether winter mulch is provided.

—▶ the real dirt *Hardiness zone maps do have their limitations.* But buying plants in your own zone will go a long way toward the development of a healthy, long-lasting garden. Still, we encourage everyone to experiment with plants that are teetering on the edge—if you don't mind gambling a bit with your money. It's fun and you just may end up with a beauty of a plant in your garden that nobody else can grow for miles around.

good advice Harden off seedlings before transplanting them outdoors

Hardening off is what you do to prepare seedlings to live in the garden. You do this by taking them from the temperature-controlled, comfortable place where you have been nurturing them for weeks and bring them outside a few hours at a time for a week or so. Doing this gradually allows seedlings to prepare themselves to face wind, rain, hot sun, cool nights—whatever Mother Nature dishes out.

WHAT HAPPENS IF YOU HARDEN OFF SEEDLINGS

If you take the time to harden off seedlings before transplanting them, they will be more likely to survive the transition from indoors to outdoors and do well in the garden.

We know many gardeners, though, who skip the hardening off process and put plants straight into the ground. Sometimes the seedlings do just fine. Other times, they struggle to make a go of things, and many eventually succumb, flopping over lifelessly onto the soil, rotting away, or just failing to thrive. It's easy to wind up with things like stunted sunflowers and skinny, wimpy herbs if you don't harden off plants.

HOW TO DO IT

Start to harden off seedlings anywhere from a week to ten days before you plan on planting them. They are usually just a few inches high at this point. Ideally, when hardening off plants you should start them in a spot that is shady or gets filtered light. Leave them out for two to three hours and bring them back inside. After a couple of days, if it's not really hot, put them in the sun for a couple of hours. Gradually increase the number of hours you leave the seedlings in the sun over time. And always bring them indoors at night. Be sure to put the seedlings in an area that is sheltered from wind, and do not let them get pelted by heavy rain. Water the seedlings sparingly to toughen them up, but don't let them wilt. Don't fertilize at this point, either.

For those who can't do what's ideal because of full-time jobs and/or limited patience for detail-oriented tasks, you can harden off seedlings in an enclosed porch that gets some filtered light, but does not get too hot during the day. You could also choose a sheltered spot outdoors that you know will only get sun for a few hours before it becomes shady. If it is hot or the plants are in small pots, they might dry out if left for eight hours a day, so it's fine to leave them in a shallow tray of water just to be safe. (Make sure they are not sitting in water day after day, though, because they will drown and rot.) If you are growing warm-season vegetables, such as tomatoes or peppers, it is still a good idea to bring seedlings inside for the night if the temperature will get below 60°F (16°C).

—● the real dirt *There's no question that hardening off helps give seedlings a good, strong start in the garden.* Gardeners who don't do it are taking a chance that the plants they worked so hard to grow will not thrive, even if they do survive. Still, we completely understand the urge to skip this process because it can be laborious at a time when you just want to get out there and garden. If you do not want to do every step of this process, even doing a shortened version of it will help.

good advice Separate the roots and remove heavily matted clumps from root-bound plants before planting

We know people sometimes feel self-conscious about sliding plants out of their containers before buying them, but it's really okay. That is the only way to find out whether the roots are in good shape or not. When you do this, you should see whitish roots that are connected well enough to form a root ball and hold the soil together.

All too often, though, what you find is very little soil and a thick, tangled mass of yellow to brown or even black roots that seem to form a thick web around the remaining dirt. More roots are likely circling the bottom of the container, too. A plant in this condition is considered root-bound (or pot-bound), and it happens when plants are left in containers for so long that roots literally have nowhere to go but around and around the edges of the container, sometimes circling it two or even three times.

As you might imagine, being root-bound harms plants in a lot of ways. Leaves yellow and sometimes drop. Water goes right through the container and out the bottom without being absorbed. Nutrients are not taken up well. Growth slows and eventually stops. In general, unless there's a great sale and you feel like taking a chance, we advise passing up root-bound plants because they have experienced a good deal of stress before you've even got them planted. If you buy a root-bound plant or if you own one that has become root-bound, you need to separate thick patches of roots and remove heavily matted clumps before planting.

WHAT HAPPENS IF YOU DON'T MANIPULATE ROOT-BOUND PLANTS

If a root-bound plant is put in the ground as is, it will be difficult for the plant to establish a good root system that will nourish the plant as it grows. At best, planting something that is severely root-bound means you will end up with a plant that just stays the same size and does not thrive for a year or two. At worst, the plant will die.

HOW TO DO IT

You may have to cut severely root-bound plants out of their pots. If that is the case, use a box cutter or sharp knife to slice the sides of the pot so you can get in there and start gently separating some of the matted roots. Use the knife to cut off some of the most heavily matted areas and even some of the roots that are circling the bottom, if that clump is thick. If there are any masses of roots that smell bad, cut these off immediately; they are rotten and could contain pathogens that might injure healthy roots. Use your hands to rough up the sides of the root ball. This will help loosen the soil and break up some of those roots so they will spread out easily once the plant is in the ground. Though removing many of its roots may seem like a bad thing to do, ultimately the plant will be healthier for it.

━━▶ the real dirt *Examine plants at the store so you can avoid buying root-bound plants.* If you buy a plant that is root-bound, remove clumped roots and straighten roots that are circling the bottom of the container.

....

good advice Use non-natives as well as natives in a landscape

Although some gardeners stridently argue for using only native plants, we think there is a place for both native and non-native plants in the home landscape. Plants that are native (or indigenous) have grown and evolved over thousands of years in a specific region, whereas non-native (or exotic) plants have been introduced through one means or another to an area. Over time, it has become increasingly common for gardeners to plant exotic plants rather than natives, and there are many reasons for this—not the least of which is that natives may be seen as common-place, whereas non-natives are perceived as new and interesting. In recent years, though, there has been a resurgence of interest in natives. For some, this fervor has grown to the point where they frown openly on the use of exotics, while lavishing praise on natives.

The rationale for this stance is often rooted in sustainability. Because native species have had millennia to adapt to the soil and climate conditions around them, they are usually thought of as drought resistant and requiring less maintenance than non-natives. Native species are also perceived as being more resistant to attacks by pests and disease, and they tend to blend in well with their natural surroundings rather than taking over the place and becoming invasive, as some exotic plants are known to do. But, as gardeners know, perceptions sometimes diverge from reality. In truth, there are many natives that are less able to tolerate drought conditions and pests than non-natives, and some native plants are considered to be invasive.

Both native and non-native plants have characteristics that endear and alienate gardeners. By generalizing that native and non-native plants are drought resistant or drought tolerant and invasive or non-invasive based on where they fall in this dichotomy, gardeners can easily set themselves up for failure. We advise gardeners to research plants and judge each one on its own merits.

WHAT HAPPENS IF YOU PLANT NATIVES AND NON-NATIVES

There is no such thing as a no-maintenance garden, though many garden gurus will tell you otherwise, and the term *low maintenance* is open to interpretation. So before you fill your garden with native plants in the hopes of cutting back on water, fertilizer, and general fussing, be aware that native plants thrive in the type of sites they evolved in. That is why it's important to try to match the conditions of their natural habitat before planting.

Also noteworthy is the fact that exotic plants frequently do better than natives, which is the reason why some non-native plants become invasive. Norway maple, purple loosestrife, Canada thistle, kudzu, and Chinese privet are all considered invasive because of their ability to deal with harsh conditions better than natives. While this makes them less than ideal for most gardens and presents a risk of invasion to nearby natural environments, it also proves wrong the idea that natives are necessarily better adapted to local conditions than are non-natives.

Many exotic species are not invasive. These plants make great additions to home gardens by creating a level of diversity that looks good and may offer year-round blooms, while providing food and habitat for insects, birds, and wildlife. Here again, though, there is debate over whether native or non-native plants offer the best resources for insects and birds. While many gardeners believe they see equal traffic at their non-native and native plants, work by Douglas Tallamy, a professor of entomology and wildlife ecology at the University of Delaware, demonstrated that native plants may be better at attracting a diversity of birds and insects. In addition, some ornithologists are concerned that the berries and seeds of some exotic plants do not provide the right kinds of fats and nutrients birds need to survive harsh winters. So birds may stick around for the bounty, but they don't survive to see spring.

One of the simplest reasons gardeners include non-natives in their landscapes is because they like them. Gardens are personal things, and if you are not putting in plants that are growing out of control and squeezing out indigenous species, we see nothing wrong with creating mixed gardens that include a wide array of native and non-native plants. Gardeners who limit themselves to natives miss out on the benefits of centuries of plant cultivation and innovation. Plants like gardenia, evergreen azaleas, Japanese maple, and Norway spruce are generally well-behaved, non-native species that look great in the garden and perform well in the right circumstances. It seems a shame to ignore them.

HOW TO DO IT

Gardens are not islands. Though we support using both natives and non-natives, no gardener should plant invasive species because wildlife and the wind could carry their seeds into nearby natural environments. Before adding something new to your garden, check with your local extension service to get a list of plants considered to be invasive in your area. If you do not have an extension service locally, the U.S. Department of Agriculture offers detailed information on invasive plants (http://plants.usda.gov/java/noxiousDriver).

When planting exotics, avoid plants touted as being exceptionally fast growers, understand in advance how the plants will spread, and remove plants that are clearly taking over. (Don't just dig up these plants and give them away. If they're taking over your yard, they'll take over someone else's, too.) Whether you are planting natives or exotics, always consider the conditions the site has to offer when choosing plants.

—▶ the real dirt *Gardening is supposed to be relaxing and fun, so we are not fond of folks who stand around, arms crossed, insisting that native plants are the only way to go.* They aren't, and they don't have to be as long as gardeners take care to plant wisely.

good advice Plant perennial beds rather than turf grass beneath trees

Grass can struggle to grow in the dense shade beneath a tree's canopy, and it often doesn't look very good. In addition, the area right next to a tree is difficult to mow because of low-hanging branches, so trees are often harmed by mowers and weed whackers. That's why it just makes sense to replace grass with perennials. The only caveat is that perennials need to be planted properly so tree roots are not damaged too much in the process.

WHAT HAPPENS IF YOU PLANT PERENNIAL BEDS BENEATH TREES

Bare dirt under a tree is few people's idea of beauty, but that is what you often get when you try to grow grass there. Perennials can do very well when planted beneath trees. Just be mindful of the amount of shade you are dealing with, and plant accordingly.

If you don't replace grass beneath trees with perennials, you can continue to live with bare dirt or grass that is barely hanging on from year to year. You can work very hard to keep the grass looking as good as possible, and it still won't look as good as the grass on the rest of the lawn. Or you could go the mulching route, which is a fine way to go because of all the good it does for the soil.

HOW TO DO IT

With nearly 90 percent of a tree's roots located in the top 3 feet of soil under and around the canopy, it's best to use only those plants that look good and can coexist without requiring too much water or nutrition. When you do dig, try to disturb tree roots as little as possible. For more in-depth information about how to plant beneath trees, go to chapter 6 and read the section that advises against raising the soil level over tree roots.

When considering what to plant, think about what normally grows in the understory. Ask personnel at a nursery or garden center if you are not sure what would work best. Some good choices include: Solomon's seal *(Polygonatum multiflorum)*, barrenwort *(Epimedium)*, foamflower *(Tiarella cordifolia)*, bleeding heart *(Dicentra spectabilis)*, *Rhododendron*, many types of ferns, and *Astilbe*.

━● the real dirt *Perennial beds not only look better than grass beneath trees, they are also a better choice for keeping trees healthy.* They will also be more sustainable in the long run because perennials require less water than grass does.

good advice Plant disease-resistant cultivars when you can

Plant breeders choose and combine specific traits like flower color or plant height to create desirable new cultivars. Disease resistance is another genetic trait that can be manipulated by breeders, and it is increasingly common to see plants labeled as disease resistant or disease tolerant. This does not mean the plant is immune to all diseases. Instead, the plant is less likely to get a particular disease or, if it does get the disease, it is unlikely to be affected as badly as a less-resistant plant. For example, some tomato varieties are labeled with the initials VF, which mean the tomato is resistant to two fungal diseases, *Verticillium* wilt and *Fusarium* wilt.

WHAT HAPPENS IF YOU PLANT DISEASE-RESISTANT CULTIVARS

Prevention is key. If you plant disease-resistant or disease-tolerant plants, you are more likely to avoid having a garden full of diseased plants or needing to use chemicals to control the problem. For example, *Phlox* and scarlet beebalm *(Monarda didyma)* are very susceptible to powdery mildew. Buy cultivars resistant to powdery mildew, and you will increase the chances of enjoying these plants without all the unsightly white splotches caused by the fungus. Because powdery mildew is not fatal, if you don't buy resistant cultivars the biggest problem you will have is ugly plants. Tomato and potato crops, however, can be lost to *Verticillium* wilt, so choosing resistant varieties might just save your backyard harvest. Be aware that planting varieties that are not resistant along with resistant varieties will increase the likelihood of the resistant varieties contracting the disease if the nonresistant ones catch it.

HOW TO DO IT

If you've had problems in your garden such as powdery mildew, black spot on roses, or sick tomatoes or other vegetables, shopping for disease-resistant or tolerant plants is a good idea. New varieties with disease resistance are introduced each year, while older ones may be phased out, so be sure to ask for help if you don't find what you're looking for.

Catalogs and plant labels often use initials to denote types of resistance. Here are some of the common ones you will see for vegetables: V for *Verticillium* wilt, F for *Fusarium* wilt, T for tobacco mosaic virus, and L for leafspot.

━● the real dirt *There are obvious advantages to choosing disease-resistant cultivars, but there are drawbacks too.* Increased resistance can sometimes result in vegetable crops that do not taste or look as good as others; for example, if a plant puts energy into a compound important to resistance, it can result in less energy being invested to produce a tasty sugar. Yields may be reduced in these cultivars, as well. Also, in some cases, breeding a plant so that it's resistant to one disease may make it more likely to succumb to another.

good advice Plant perennials in spring and early autumn

The best time to plant perennials varies from species to species. For many plants, it doesn't matter when they are planted, but others are pickier. Spring-flowering bulbs, for example, need to be planted in autumn. Generally speaking, though, spring and autumn are thought to be the best times to plant perennials because the cooler temperatures give plants that are struggling to develop a good root system a more favorable environment. Otherwise, plants have to develop roots while expending much of their energy blooming in the heat of the summer.

WHAT HAPPENS IF YOU PLANT PERENNIALS IN SPRING AND EARLY AUTUMN

In general, the availability of container-grown perennials allows gardeners to plant whatever they like, whenever they like. But, in some regions, getting plants in the ground before the real heat and humidity kick in does make a plant's transition to a new home less stressful. Planting at cooler times of the year is likely to save water, as well.

If you plant perennials in the heat of the summer, they will definitely experience stress as they work to develop a root system under difficult conditions. Keeping them well watered will reduce the number of casualties. If water is lacking, you may lose some plants, particularly larger ones (gallon-size containers or larger) because it takes more work for a plant with a fairly extensive root system to establish itself.

If plants are in bloom, transitioning is even more difficult because so much energy is going into those flowers when the focus really needs to be on the roots. This is why, no matter what time of year it is, we prefer to plant smaller perennials, commonly those in 4-inch pots. These smaller plants definitely have less visual impact immediately after installation, but they catch up fast, cost less, and seem to experience much less stress when planted.

HOW TO DO IT

Regardless of when you plant perennials, amend the soil with good-quality organic material, such as compost, at planting time. In the first two to three weeks after planting, be sure perennials are watered regularly, but not so much that they get soggy. This will help roots establish themselves.

If you plant after the weather heats up, try to do so on an overcast day, so the blazing sun will not be such a stress on plants right away. If that's not possible, do your best to water regularly (perhaps once every two to three days), and don't let wilting be your guide. By the time plants have wilted, they are stressed.

—● the real dirt *Spring-blooming bulbs should be planted in autumn. When possible, plant other perennials in the spring.* But if you find you need to plant in the summer heat, don't worry too much. Chances are that if you water and mulch properly, the plants will be just fine. As gardeners in Minnesota, we are not big fans of autumn planting. Our winters can be almost unimaginably harsh, and too often we have lost plants because they did not become established well enough before the cold set in. If you do plant in autumn, be sure to do so at least four to six weeks before a hard freeze. In regions with more mild winters, a good rule of thumb is to plant about four weeks before a hard freeze to get those roots established.

advice that's debatable Follow spacing recommendations on plant labels

In our opinion, plant spacing has more to do with design preferences than an actual need for precisely measuring the area that a plant needs to grow. Talk to horticulturists, gardeners, growers, or landscape designers and you are sure to get a wide range of opinions about what the optimal space between certain plants should be. Some will say with authority that pansies and sweet alyssum should be 6 inches apart, whereas geraniums should be 12 inches away from the next plant. Others prefer using mathematical formulas to calculate the number of plants needed for a given area. The latter approach is handy for landscape designers and architects, but there's something to be said for simply laying out plants on site with an eye toward the finished design. Do you like plants to be close together, or do you like to see a bit of dirt around each one? Are you trying to create an archway or hedge with trees or shrubs, or do you want a neat row?

Annuals and perennials are often squeezed closely together and then removed as the plants grow and invade each other's space. For trees, and for vegetables for which you want to maximize yield, it is important to consider the spacing between plants more carefully because you don't want to have wasted your money purchasing trees or tomatoes that will need to be removed as they grow into their maximum size. Do your research to find out their maximum spread when they are mature.

WHAT HAPPENS IF YOU FOLLOW SPACING RECOMMENDATIONS

If you adhere to the spacing recommendations found on plant labels, you will save a little money by not buying so much at one time and your garden will probably not be overcrowded in just a few years. These are good things because you will be able to relax and think about what to do with your savings. Gardeners who wanted a lush look more quickly—or just could not pass up plants they loved—will be out digging up plants and moving them around or giving them to friends and neighbors.

If you put too many plants into the garden from the start, you will probably have to do some work to thin them out in the near future. There is also a higher likelihood of disease when plants are placed too closely together, because air circulation is reduced.

A BETTER WAY

There are many plant-spacing charts available, as well as formulas for calculating the distance between plants of varying sizes. Which one you choose to follow is up to you. We recommend following none of them, for two reasons. First, to us, thinking about gardening as a big math problem doesn't seem like much fun. Second, even the most exacting spacing plan can go awry when plants don't behave as they are supposed to by not growing wide or tall enough.

Instead of following charts that may or may not be appropriate to your conditions, take time to understand what each plant is supposed to look like when it matures, so you can envision what your garden will look like in a few years. Before you dig, place the potted plants where you think you want them, and move them around to create a good composition that includes plenty of contrasting leaf color, texture, and height.

Gardens are always a work in progress, so you will have to move some plants around as they grow or once you realize some plants don't look right where you put them. If you are the rare gardener who does not crowd plants too close together from the start, but you wish there wasn't so much dirt showing, try adding annuals to fill the spaces for a year or two.

—● the real dirt *Meleah is a gardener who designs by muttering to herself, plant in hand, as she walks around looking for the right place to put it.* She cannot imagine following charts or using calculations to figure out plant spacing, and neither can most of the gardeners she knows. But they are also fine with having to move plants around a lot when they get too crowded. If you prefer a more laid-back approach to gardening, paying closer attention to spacing guidelines will be helpful. But it is not imperative to your gardening success.

advice that's debatable Use expensive grow lights to start plants successfully from seed indoors

If you love plants, but don't have a lot of money, growing some from seed is the way to go. Even if money is no object, starting plants from seed allows you to have a much wider range of plants in your garden than you could otherwise. Maybe you've seen all the seed starting gear on garden center shelves and in catalogs, and you're wondering if you really need it. Honestly, you don't need half that stuff to get plants growing—and you definitely don't need expensive grow lights. Although these lights can offer energy to young plants to get them started, and they certainly should not be considered a rip-off, a simple setup of cool fluorescent tubes that can be bought at any hardware store will work just fine for most plants.

WHAT HAPPENS IF YOU USE GROW LIGHTS TO START PLANTS

Once seeds germinate and green sprouts emerge from the soil, the seedlings need light in order to grow—their ability to photosynthesize depends on it. Several different types of grow lights are recommended for plants: fluorescent, LEDs, incandescent, and high-intensity discharge lights.

Professional growers, particularly of flowers and vegetables, often favor the high-intensity discharge lights, which emit light that is more similar to that coming from the sun than other lights. Professional nurseries, though, not only need to start seedlings, they need to grow plants to a more advanced stage of maturity, perhaps to the point of flowering or fruiting, before selling them. We home gardeners just need to get our seedlings started so we can transplant them outdoors.

Some garden gurus suggest buying cool and warm fluorescent lights so plants get the same balance of blue (cool) and red (warm) light they would from the sun. In our experience, these warm lights are not necessary for starting seeds. If you go the inexpensive route and buy cool fluorescent lights rather

123

than pricey grow lights of another type, your plants will still do well. Young plants will also do fine with LED lights, if you choose to go that route.

Do not go so low tech that you rely on window light alone, as some suggest. Seeds will probably germinate in a window, but even in a very bright spot, seedlings are likely to end up being tall, spindly, and unprepared for the harsh outdoors.

With any of these lights, it is a good idea to harden off plants before they are planted outside, as described previously. None of these lights has quite the energy that the sun does, and you don't want seedlings shocked when they go outside.

A BETTER WAY

You can buy all kinds of shelving and lighting systems for starting seeds. Or you can make something simple yourself, like the setup Meleah has in her basement.

Sow seeds in biodegradable pots according to the package directions to ensure enough time for germination and hardening off before planting outdoors. The pots can be lined up on old cookie sheets or inside plastic plant flats. (If there are holes in the flats, be sure to have something underneath to catch any water overflow.) Once everything is assembled, slides the pans or flats onto the shelves of an inexpensive metal shelving unit (the kind you often see in garages). Above each shelf, hang a shop light fixture that holds two fluorescent tubes on chains, which make it easy to raise the lights as the plants grow. The lights should remain about 2 to 4 inches above the top of the seedlings. Once the seedlings sprout, give them about 14 to 16 hours of light a day, which can be easily done by putting the lights on a timer. In general plants will not grow well if exposed to light 24 hours a day. They have certain physiological processes that go on in the dark, and without a dark period these processes will not occur. In a few weeks the seedlings will be sturdy enough to plant in the garden.

━● **the real dirt** *Unless you are a commercial producer, inexpensive fluorescent lights will be just fine for starting most seeds indoors.*

advice that's debatable Do not place plants labeled full sun in spots that get less light

Ask half a dozen gardeners to define full sun, and you might just get six different answers. We define full sun as six or more hours of unfiltered sun during the prime time of the day, which is between 9:00 a.m. and 5:00 p.m. This six-hour minimum for full sun is not cumulative, so you cannot add up three hours in the morning and another three hours in the evening. Full sun needs to be consecutive hours that ideally include much of the time when the sun is highest in the sky. Sun at earlier and later hours is great, but it's less intense so it does not offer the blasting heat some plants need to flower well. Plants that receive less sun than they need can get leggy and flop over.

Meleah has managed to grow coneflower *(Echinacea purpurea)*, black-eyed Susan *(Rudbeckia fulgida)*, blazing star *(Liatris)*, *Aster*, scarlet beebalm *(Monarda didyma)*, tickseed *(Coreopsis)*, joe pye weed *(Eupatorium purpureum)*, *Clematis*, Asiatic lilies, and Siberian iris *(Iris sibirica)* in less than optimal light conditions. When she compares her garden, which gets full sun from about 11:00 a.m. to 2:30 p.m. and filtered sun thereafter, to neighbors' gardens that get much more light, it's clear that some of her plants are shorter and some have fewer flowers than theirs. But all the plants look healthy, and the blooms provide a burst of color in an otherwise partially shady yard.

WHAT HAPPENS IF YOU DON'T CONSIDER LIGHT LEVELS

So, what happens if plants labeled full sun do not get six hours of intense sunlight? Maybe nothing. It depends on the plant. If you've chosen a plant at a garden center that you are unfamiliar with, it's a good idea to talk with the staff about the light in your garden and whether the plant will do well. If they don't know and the plant is inexpensive, you may want to give it a try in your garden and see how things go. Unless you have money to burn, though, we would not recommend trying this with a $200 tree that you may watch die over a couple of years and then have to remove it.

If you buy full-sun plants and put them wherever you like without regard to light, there's a good chance they won't grow as well as they should and they may be more prone to pest and disease problems. If you plant lilacs *(Syringa)* in the shade, for example, they will not flower well and they will be more apt to develop problems like powdery mildew. But black-eyed Susan *(Rudbeckia fulgida)* and coneflower *(Echinacea purpurea)* are said to need full sun, and yet it's common to see them thriving in partial shade. If the only consequence for giving certain plants less than optimal sun is that they flower less or don't get as bushy as they might have, give them a try.

It would be a mistake, though, to try to grow sun-loving vegetables like tomatoes and peppers in anything but full sun. We will have more to say about this in the upcoming section on vegetables.

A BETTER WAY

When you are home some weekend, study your garden to see what kind of light it really gets. You might be surprised by how the light changes throughout the day.

If you try a plant in a certain spot and it winds up looking stunted, having fewer leaves than it should, getting leggy, or not flowering well, you will know it needs more sun than it is getting. Likewise, if a plant labeled full sun is wilting in the heat of the day or the leaves are getting scorched, it's probably time to move it to a new place where it gets a little more shade. Plants will do what they are going to do, regardless of the labels on their pots.

—▬ the real dirt *Even if you don't have full sun any-where in your garden, you still may be able to grow plenty of so-called full-sun plants.*

advice that's debatable Wash and sterilize containers at the end of the season

The reason gardeners are advised to clean pots and other containers every year is to prevent passing harmful bacteria, fungi, and viruses from plant to plant. Advice about how best to clean containers is everywhere, and tips vary depending on the source. The easiest method is simply to remove loose dirt, roots and other debris, wipe the container down with a rag, and store it for the winter. The problem with this method, though, is the lack of sterilization. Cleaning alone will not kill all of the pathogens that cause disease. To do that you need to use heat or mixtures of bleach or vinegar and water.

WHAT HAPPENS IF YOU WASH AND STERILIZE CONTAINERS

Cleaning and sterilizing containers does get rid of pathogens that might hang around and contaminate plants grown in the same pot year after year. You may bring a plant home, put it in an old dirty pot, and then watch it come down with something, so you have to throw it out. Of course, any plant you buy could be harboring a disease before you even pot it up. But if you want to be sure you are not putting a perfectly good plant into a hotbed of hungry pathogens, cleaning and sterilizing pots is the way to go. This is particularly important when you are working with seedlings, as they can easily be wiped out by fungus. Always start with clean, sterile pots and planting medium when you are starting plants from seed.

A BETTER WAY

Cleaning involves simply removing dirt and scrubbing. Garden experts recommend several different methods for sterilizing pots, but some methods are not suitable for all types of pots. For example, metal and clay pots can be sterilized in boiling water, but plastic pots cannot.

Terra cotta pots can be baked in a 220°F (104°C) oven to kill pathogens. Place the pots on a baking sheet and leave them in the oven for one hour. Turn off the heat and let the pots cool completely before removing them.

Any sort of pot can be sterilized with a mixture of 1 part bleach and 10 parts water. Soaking rather than scrubbing is okay, too. Rinse well, and let the pot dry completely before planting. If you don't like bleach, you can soak or scrub pots in a solution of 1 part vinegar and 1 part water. Some gardeners like to run smaller pots through the dishwasher, whereas others just clean dirty containers with soapy water. This works, but it's not as effective at killing pathogens as other methods. Be sure to completely wash the vinegar or bleach from the container before you add potting soil, as these compounds can be toxic to plants as well as microbes

—• the real dirt *For established plants, using old, unsterilized pots is not that big of a deal as long as the plant previously in the container did not have a disease.* Seedlings should always be started in fresh potting medium in sterilized pots.

advice that's debatable Divide plants and transplant only in the spring and autumn

Shortly after spring has started and summer is approaching people start asking, "Is it too late to move my plants?" Having missed the proper window of opportunity, they are afraid that if they dig something up anyway they will likely kill it. Maybe it's this fear of certain death that makes this advice strike fear into gardeners' hearts in a way other advice doesn't seem to. Whatever the case, we hate that people feel this way. Experienced gardeners move plants whenever they feel the need. Professional landscapers install plants from spring to autumn.

It is true that most perennials do best when they are divided in early spring as new growth begins to emerge. But that's not always the time when gardeners can get outside and work. It's also difficult to know what something is when all you can see is a sprig of it in the spring. Most plants can be moved any time (except during winter when the ground is frozen). If you move them during a stressful time, like the heat of the summer, you just have to take extra care of them.

WHAT HAPPENS IF YOU ONLY MOVE PLANTS IN SPRING AND AUTUMN

By moving plants in spring and autumn, you won't have to fuss over them as much as you would if you moved them when conditions were less favorable.

There are many good reasons for moving and dividing plants at all times during the year. Maybe a plant is in the wrong spot or you have several mature plants that are clearly in need of division. You can tell it's time to divide a plant when it produces fewer blooms and has slower growth. A lot of perennials, such as a sedum or iris, die out from the center because the roots there get overcrowded and cannot get the water and nutrients they need. This means they will not only look bad (with a telltale bare ring in the middle) and bloom less, they will be stressed and could become more susceptible to diseases and pests. Whatever the reason for transplanting, true gardeners are never done with their landscapes and moving plants all over the place is a big part of what we do.

A BETTER WAY

Perennials are usually divided in one of two ways: dig up the entire clump of whatever you are working with and then start to separate it, or use a shovel or spade to cut out a portion of the plant. If you do the latter, the plant may look maimed for a little while, but it will fill out quickly.

If you are dealing with a doughnut hole in the middle of a perennial, it's best to dig up the whole plant and then use a sharp knife to cut out the parts of the root system that look dead. Sometimes it's also possible to just remove the roots by hand. Once you've gotten rid of the old, dead roots, divide the plant into several sections and replant them around your garden, being mindful to choose spots that offer the right amount of sun for the plant.

When dividing and transplanting, always start by cutting plant foliage back to about half. The transplant's smaller root system will not have to struggle so much to support a lot of greenery or blooms. It may be hard to do, particularly if a plant is flowering, but this is particularly important when transplanting in the hot summer months. Likewise, try to divide and transplant in the early morning, on cloudy days, or in the evening so plants will not have to endure the sun and heat of the day when they are most fragile. If you do transplant in the middle of a summer day, try to provide some shade for a week or so for transplants that seem particularly fragile, such as coral bells *(Heuchera),* ferns, and bleeding hearts *(Dicentra spectabilis).* An old sheet draped over a couple of chairs will work if you have to improvise a shade structure.

Water plants right after dividing and transplanting, and keep them well watered until they seem to be holding their own without flopping over, wilting, or looking bedraggled and sad. This may take a week or two, and you might have to water twice a day if it's really hot outside. To give plants time to establish a good root system before winter comes, finish all of your dividing and transplanting four to six weeks before the ground freezes.

━● the real dirt *We both move plants from the time the ground can be worked in the spring into the autumn.* Yes, it is clearly harder on plants to be moved in the heat of the summer, but if you water well and provide a little shade for the really stressed ones, you should have few problems. Heed advice to stop transplanting about six weeks before the ground freezes, though, or you may lose plants to the cold.

ANNUALS, PERENNIALS, AND BULBS

advice that's just wrong
Add phosphorus to increase bloom and stimulate rooting

Lots of fertilizers that contain large amounts of phosphorus tout their ability to increase bloom and stimulate root growth. But researchers have discovered that there is not much evidence to support these claims. Yes, plants need a certain amount of phosphorus, but if you add too much you'll just be polluting the environment.

WHAT HAPPENS IF YOU ADD PHOSPHORUS

The nitrogen (N), phosphorus (P), and potassium (K) ratio listed on fertilizer labels, such as 10–10–10, is the percentage of each nutrient in that particular product. In the case of 10–10–10, it is a balanced fertilizer, because the amount of each nutrient in the mix is the same. Fertilizers that are said to promote root growth and bloom often have a ratio that is more like 10–40–10 or even 10–50–10.

Is all this phosphorus doing what it's supposed to do? Jeff and other scientists have tested whether added phosphorus increases root growth by adding large amounts of phosphorus to some containerized plants while adding none to others. The results showed no measurable difference between the plants' roots. If adding phosphorus does not help with root growth, does is help with bloom? Again, researchers at various universities have found that plants require just a small amount of phosphorus to flower well. Because a sufficient amount of phosphorus is usually found in healthy, unfertilized soil, there is no need to add more unless a soil test reveals a deficiency.

A BETTER WAY

Whether they are grown in containers or in the ground, if plants are put into good soil that includes plenty of organic matter, they do not need a lot of fertilizer. For situation when you do need a fertilizer, an N–P–K ratio of 5–1–2 is appropriate for most situations. If your plants are struggling, we recommend a soil test to see if you need to add nutrients of any kind.

More often, though, you will find you need to do more or less watering to keep plants healthy.

—➧ the real dirt *Unless you have a phosphorus deficiency and can prove it with a soil test, stay away from high-phosphorus fertilizers.* In general you do not need them, and they can cause severe environmental damage if they get into lakes and streams.

6

TREES and SHRUBS

In This Chapter

Good Advice

∘ Do not divide shrubs

∘ Prune trees to promote good growth and health

∘ Do not plant trees too deeply

∘ Plant trees in holes two to three times the diameter of the root ball

∘ Do not raise the soil level over tree roots to plant a garden

Advice That's Debatable

∘ Fertilize trees at planting time

∘ Always stake young trees

∘ Use tree wraps to protect young trees from sunscald and insect pests

∘ Wrap evergreens in burlap for winter protection

∘ Remove twine, burlap, and wire before planting a balled-and-burlapped tree or shrub

Advice That's Just Wrong

∘ Top trees to keep them from getting too tall

∘ After pruning, dress tree wounds to inhibit decay and insect infestation

∘ Prune trees when they are planted

∘ Beat a tree to get it to flower

∘ Make pruning cuts flush with the trunk

∘ Shear the tops of deciduous shrubs to control their height

In his smart and lovingly written book The Tree: A Natural History of What Trees Are, How They Live, and Why They Matter, *Colin Tudge poses the question "What is a tree?"* He answers, "A tree is a big plant with a stick up the middle." How right he is, but what strikes us about his answer is its simplicity and how it fits our perception of trees. Perhaps it is their size that makes trees seem less vulnerable and more straightforward and understandable than other plants. Unfortunately, this perception has not served them well.

Researchers, particularly in recent years, have been discovering many ways in which we have failed trees. One of the most recent discoveries is that we have been planting trees too deeply. Meaning well, we have advocated improper techniques like painting wounds with tar or paint, wrapping evergreens in burlap, and staking young trees (which is sometimes okay). This chapter, perhaps more than any other in the book, offers a fresh, research-based perspective on things we gardeners thought we knew but, in fact, we didn't.

good advice Do not divide shrubs

People often ask us about dividing shrubs, and we always tell them the same thing: don't do it. Well, mostly don't do it. Shrubs are different than perennials in that they are woody plants. They do not die back to the ground once the growing season ends. When you divide a perennial, its root system repairs itself and the plant keeps growing. But shrubs have woody root systems that don't do well when divided. If you hack a shrub in half or into several pieces, those pieces will likely die within the first season.

WHAT HAPPENS IF YOU DIVIDE SHRUBS

It's easy to see why people think they can divide shrubs. Like perennials, they get big and unruly so we get the urge to split them up. But even if you do manage to divide a shrub and have its various segments survive, it will likely end up looking weird because parts will not regenerate like they should. It will never develop a natural-looking shape.

Pruning shrubs that have become unruly and overgrown is a much better strategy than hacking them up. If you want more shrubs, buy them as you can afford them.

HOW TO DO IT

There are some exceptions to the don't-divide-shrubs rule. Those shrubs that produce offshoots, such as common lilac *(Syringa vulgaris)* and *Spiraea,* can be divided in the sense that you can dig up the shoots and replant them elsewhere. This can still lead to oddly shaped shrubs, but it usually works well enough. If you really want more shrubs but do not want to buy them, your best bet is to propagate them from cuttings. (For more on propagation, see Ken Druse's *Making More Plants: The Science and Art of Propagation.*)

—• the real dirt *Though it might seem like a reasonable idea, we do not recommend diving shrubs in most cases.* If they survive at all, you will just wind up with ugly, weirdly shaped shrubs.

good advice Prune trees to promote good growth and health

Pruning, especially when it comes to trees and shrubs, is one of those jobs gardeners often find most daunting. We are asked constantly about how to prune, when to prune, and how much to prune. There are some fundamentals to learn. But once you get those, pruning becomes second nature, which is a good thing because trees and shrubs do need to be pruned to grow well and stay healthy.

WHAT HAPPENS IF YOU PRUNE TREES

If plants don't get pruned in nature, why should you prune them at home? Good question. Pruning is important for many reasons. It keeps plants looking their best, and stops them from blocking walkways and windows or crowding other plants around them. Pruning keeps plants healthy by removing dead and diseased wood and preventing overcrowded branches that impair light and air circulation. Common objectives for pruning are removing dead, injured, or diseased braches and branches that cross over each other or rub together; renewing older shrubs by removing up to a third of the old wood; shaping trees, shrubs, and hedges; creating a single leader; removing branches that are at strange angles and may be poorly attached to the tree; and getting rid of suckers and water sprouts (limbs that grow rapidly upward). Some plants, such as conifers, do not usually need regular pruning.

If you opt to forgo pruning, many trees and shrubs will wind up looking ugly, with lots of dead branches. Unpruned trees and shrubs are more prone to disease problems than those pruned regularly. Additionally, trees can be hazardous if branches are at poor angles or densely packed branches don't allow wind to flow through the canopy easily, making them more likely to topple in a windstorm.

HOW TO DO IT

Depending on the type of plant being pruned, there are many things to consider before tackling the job: what tool to use (pruner, knife, trimmer, saw), time of year, and desired shape, to name a few. That sounds complicated, but it really isn't. Yes, different trees and shrubs may require their own pruning techniques, but most gardeners only have a handful of plants in their landscape that need pruning.

Pruning is usually best done while the tree is dormant, over the winter or in the very early spring. If needed, pruning can be done at other times, but this should only be for necessities, such as removing a dead limb.

It's a good idea to look at some illustrations of proper pruning before starting, so you can understand what it means to prune a tree limb at a branch collar rather than leaving a stub. Books are great for this sort of thing, of course, but you can also find plenty of good images of pruning online. One of our favorite pruning sites is hosted by the University of Florida (http://hort.ufl.edu/woody/pruning.shtml).

When you prune shrubs, make the cut about ¼ inch above a bud or, if you want to take more off, go all the way back to a side branch. New growth will go in the direction the bud is facing so keep that in mind when pruning. If you have large tree limbs that you think should be removed, it's best stay safe and hire a tree care company to do the job. It's also best to hire a tree care company if you want to remove limbs that are higher than you can easily reach.

—● the real dirt *Pruning can be skipped, but we do not advise it.* Not only do plants look better when they are pruned, they are healthier and more productive, too.

good advice Do not plant trees too deeply

Most gardeners know that you should place plants in the ground at the same depth they are planted in their containers, rather than plopping them way down in the bottom of a hole. But recent research has shown that we all (even arborists) have been planting trees too deeply for years and years. Look around at trees in your neighborhood, maybe even your yard. Does the trunk look like a telephone pole sticking out of the ground without an obvious root flare where the tree meets the earth? If it does, it's buried too deeply.

WHAT HAPPENS IF YOU PLANT TREES TOO DEEPLY

Burying trees deeply used to be considered a good idea because it was thought to protect young trees from being blown over by the wind and to protect the roots. But this thinking is wrong.

Ongoing research on the effects of planting trees at various depths, including a nine-year study by the University of Minnesota, have shown that it is best to plant trees so the root flare (or root collar) is at, or just above, the soil. The root flare is where the main stem transitions to the roots at the base of the trunk. When trees are planted this way, rather than deep in the ground, their roots are close to the soil surface where they can easily get the air, water, and nutrients the tree needs. Although it may take 10 to 20 years, when trees are planted too deeply their roots will grow upward in an attempt to get closer to the soil surface where they can find more air. As they grow, these roots encircle the base of the tree. Eventually, they girdle the tree, essentially strangling it as if it had a noose around its neck.

HOW TO DO IT

Spreading the word about how to plant trees correctly should be at the top of every gardener's to-do list because this problem is still widespread. One reason the problem persists is that most trees purchased from nurseries are still being planted far too deeply in their containers. Most homeowners don't realize this, so they just plant trees as they would any other potted plant.

Before you plant a tree, gently remove it from its container and use your hands or a trowel to scrape away the top layer of soil until you expose the root flare. Because balled-and-burlapped trees are usually bigger than containerized trees and have loose root balls, use wire (such as a straightened coat hanger) as a probe to help you find the root flare. Keep the tree wrap on while you do this. Once you have established how far down you will need to go to get to the flare, subtract that number of inches from the depth of the planting hole you will be digging. Believe it or not, depending on the size of the tree and the nursery that planted it, you may have to dig down as far as 12 inches. Once the tree is in the hole, remove the soil from the top, taking care that the flare can be seen aboveground once you are finished. No telephone poles!

—▶ the real dirt *It's sad that so many trees will lead shortened lives because we planted them incorrectly.* But now that we know how to plant trees to promote their long-term survival, we need to spread the word as much as we can. This includes talking to managers at garden centers and the owners of nurseries when we see trees in containers potted too deeply.

good advice Plant trees in holes two to three times the diameter of the root ball

When it comes to tree planting, one of the biggest keys to success is proper site preparation. Taking time to dig a large planting hole, about two to three times the diameter of the root ball, will go a long way toward ensuring a tree's future health because it will allow for better root growth even in poor soil.

WHAT HAPPENS IF YOU PLANT TREES IN LARGE HOLES

The bigger the hole, the better, because the roots of newly planted trees will spread out faster in loosened soil, making it easier to take up needed water and nutrients. In most cases, holes only slightly larger than the size of the container or root ball going into the hole will do just fine, but when forced to establish themselves in dense, compact soil, the roots of newly planted trees may not grow and spread as needed to ensure the trees' health.

HOW TO DO IT

Before you dig a hole, call your local utility companies to find out if there are any buried lines in the area you will be working in. Then dig a hole two or preferably three times the diameter of the tree's root ball. Make sure that the depth of the hole is just deep enough so that the tree's root flare is at the soil surface. As we mentioned in chapter 2, adding some water to the hole is a good idea. When you are finished, roughen the sides and bottom of the hole with the shovel to help roots penetrate the soil more easily. Do not loosen the soil around the root ball, as this might tear young roots.

—● the real dirt *Larger planting holes allow better root growth, so taking time to dig the correct size hole for a tree will be worth it in the long run.*

good advice Do not raise the soil level over tree roots to plant a garden

Grass does not grow well under most trees. After struggling to keep it alive or looking at least halfway decent without success, people often ask us if it's okay to plant a garden under their tree instead. We tell them yes. But when we explain that they have to do this without adding soil that will cover the tree's roots, they get a deflated look and want to know how they are supposed to create a garden without adding soil. Our answer is "very carefully."

While trees do have roots that help anchor them to the earth, the roots they rely on most to provide air, water, and nutrients are just inches below the soil surface in an area that expands well beyond the drip line, or canopy edge. Cover these roots with soil and you cut off the oxygen the tree's roots need to survive. First the trees will suffer, and then they may die. This is especially true of shallow-rooted trees, like maples *(Acer)* and elms *(Ulmus),* which have roots so close to the soil surface that many roots can be seen fanning out around the base of the tree above the ground.

WHAT HAPPENS IF YOU RAISE THE SOIL LEVEL OVER TREE ROOTS

Working to create a garden under a tree without raising the soil level is the best way to ensure the health of both the garden and the tree. If you decide to raise the level of the soil, you will be putting your trees under a lot of stress that they may not be able to handle.

HOW TO DO IT

If you want to plant under an established tree without increasing the depth of the soil over the roots, your goal is to disturb the soil as little as possible. First, if you have grass, you need to get rid of it. The easiest way to do this without harming tree roots is to lay a few sheets of wet newspaper down over the grass and top it with a 2-inch layer of organic mulch, all of which you will remove after the grass dies. Unfortunately, it may take a couple of months for the grass to die. If you want to plant faster than that, you can spray the grass with chemicals, such as glyphosate. Just be sure not to spray any part of the tree, including the roots, when you are doing this as it could badly damage the tree. Be especially careful if the tree roots are at the surface of the soil. You can plant once the grass has died. If you used glyphosate, you need to wait three to five days for it to become deactivated in the soil.

When you are digging, cutting into some tree roots will be unavoidable. But when you come across a root that is bigger than 1½ inches in diameter, do not slice into it to make room for a plant. Instead, just move over to a spot with smaller roots. You may wind up with a garden that is not perfectly spaced, but you will preserve the health of your tree.

Choose the smallest plants possible (ideally in 2- to 4-inch pots), so when you dig the holes you will be doing the least amount of damage. It's also best to choose perennials rather than annuals. That way, you do not have to disturb the soil around tree roots over and over again. Be sure to select plants that do well in dry shade, too. Even when it rains plants tucked under trees do not get much of a drink, and they will always have to compete with tree roots for water and nutrients.

Spring-flowering bulbs and spring ephemerals, such as *Trillium* and Virginia bluebells *(Mertensia virginica),* are good choices for planting under trees because they can get plenty of sun and rainfall before trees leaf out for the season (unless you're planting under an evergreen, of course). Don't forget shrubs when you are planting under trees. Ninebark *(Physocarpus opulifolius),* snowberry *(Symphoricarpos albus),* and winterberry *(Ilex verticillata)* are just a few of the shrubs that do well in dry, shady conditions. Always be mindful, though, of how wide and tall plants will get when they mature because you do not want them to wind up looking misshapen or crowded.

Once all the plants are in, water well and cover the area with about 2 inches of wood chips or shredded wood mulch to help conserve moisture. Keep your new garden watered well through its first year, when it will be tough for those little plants to compete for what they need.

━● **the real dirt** *Planting shrubs and perennials under established trees is a fine idea. But do not raise the soil level.* Although it may look so much better if you did and it would be easier to plant with more dirt, this is a bad idea for the long-term health of your tree.

advice that's debatable
Fertilize trees at planting time

Like all plants, trees need nutrients in order to grow and thrive. If the necessary nutrients are not available, trees will be more susceptible to insect and disease problems. They may also lead shorter lives or fail to develop to their full potential. The nutrients plants need can be divided into two groups: macronutrients and micronutrients. Macronutrients, including nitrogen, phosphorus, and potassium, are used by plants the most. But micronutrients, such as manganese, iron, and zinc, are also important.

Typically trees do not take up many nutrients right after they are planted because their root systems are not yet established. The larger and more mature the tree is at planting time, the longer it will take before it starts taking up nutrients. If you add fertilizer to the planting hole when a tree cannot use it, you run the risk of doing much more harm than good. Fertilizers contain salt, which can only exacerbate transplant shock and may also burn the roots.

WHAT HAPPENS IF YOU FERTILIZE TREES AT PLANTING TIME

If you have a small tree, less than 6-feet tall, say, there may be some benefit to fertilizing at planting time because an immature tree will establish itself quickly and begin to utilize nutrients in the soil. But if you have a larger, more mature tree the benefits of fertilizing are much less clear. If the soil where you are planting has a clear nutrient deficiency, some slow-release fertilizer to correct the deficiency may be appropriate.

Trees planted in reasonably good soil do fine without fertilizer at planting time. A soil test is the best indicator of whether nutrients are needed for a newly planted tree, along with keeping an eye on other trees in the area to see whether they seem to be suffering at all.

A BETTER WAY

If you feel you need to fertilize a newly planted tree, we would advise getting a soil test first because there's no need to pour on fertilizer when none is needed. If you discover the soil is significantly devoid of nutri-ents, opt for a slow-release fertilizer and apply it according to the in-structions on the label around the planted tree, not in the planting hole.

—➤ the real dirt *It's best not to fertilize trees at the time of planting, unless the soil has a deficiency shown by a soil test.* Generally, fertilizers should not be applied to trees until at least a year after planting.

advice that's debatable Always stake young trees

Not too many years ago, it was common practice to recommend tree staking. But experts now often disagree with this advice, arguing that most newly planted trees do better when they are not staked. The reason? When young trees are allowed to move freely in the wind, they are better able to develop strong, tapered trunks, which are important for stability. Although staking may lead to straighter trees in the short term, it can make trees unstable and less resilient over time. Staking can also cause damage to the trunk when it's done incorrectly. Leaving stakes and supports on too long can cause all kinds of problems, including the tree growing around its own support system.

WHAT HAPPENS IF YOU STAKE YOUNG TREES
Trees planted in very windy or overly exposed locations and trees that have naturally flexible stems, such as many young oaks, can get some benefit from short-term staking (for about six months) to keep them grow-ing straight early in their lives

A BETTER WAY
If you do stake a tree, it's best to do so for only about six months or, in a worst-case scenario, a year. Staking for longer than a year is very rarely a good idea because a tree can become dependent on the stakes for support. Properly staked trees should have enough room to move

freely with the wind, but not be bent by it. At least two stakes should be used per tree, and each should be an equal distance from the trunk. These should be placed far enough away from the trunk so that wind won't push the tree into a stake and injure it. Though you often see wire or wire encased in rubber used to secure a tree to its stakes, it is best to use soft material, such as burlap or strips of other thick durable cloth, that will not cut into the trunk and harm the tree. Bury stakes at least 2 feet deep to provide adequate support.

➡ the real dirt *Tree staking used to be common practice.* We agree with experts who recommend staking only in windy locations or in the case of young trees with excessively flexible stems.

advice that's debatable Use tree wraps to protect young trees from sunscald and insect pests

Tree guards are physical barriers that surround the base of a tree to prevent it from being damaged by a range of different things, like intense winter sun or errant weed whackers. In the last several years, some researchers have found that certain types of tree wraps may not provide the level of protection they were thought to offer, particularly when it comes to winter protection. Wraps composed of burlap and paper, in particular, are less likely to protect a tree. Wraps made of plastic may protect the tree over the winter, but if they aren't removed the tree can grow into the wrap, potentially strangling itself and at the very least becoming unsightly. It has also been shown that, rather than protecting trees from harm, some guards become homes to insects and critters that work out of sight on the bark to the point where trees are damaged severely or even killed. For now, professionals are actively debating the benefits and drawbacks of tree wraps.

WHAT HAPPENS IF YOU USE TREE WRAPS

The most obvious benefit to wrapping trees is that it will prevent poorly steered lawnmowers and loosely held weed whackers from doing as much damage as they otherwise might if no protection were available. A less obvious benefit is protection from sunscald.

Sunscald is often referred to as southwest injury because damage usually occurs on the southwestern side of young tree trunks. It happens on cold winter days when the sun heats up tree bark (this is made worse by light reflecting off of snow). The warm bark is then cooled as clouds move in or night falls and temperatures drop. When repeatedly subjected to this intense heating and cooling, young trees—particularly those with thin bark, such as crabapple, maple, cherry and honey locust—can develop elongated, sunken, or cracked areas of dead bark where the tissue has been killed. Older trees have thicker bark that insulates tissue from the sun's heat, so they are less susceptible to sunscald. Tree wraps are said to help control this damage by mitigating the temperatures around the tree. They do this by reflecting sunlight and, depending on the wrap, providing insulation for the tree's trunk.

It's possible and even likely that nothing will happen if wraps or guards are not used to protect the bark of young trees. But we think wraps are a preventative measure worth considering if you have thin-barked trees, which may require protection for several years.

A BETTER WAY

One of the biggest reasons experts advise against using tree wraps and guards is that many homeowners do not remove them in the spring. When left on once the weather warms up, these coverings on trunks can trap moisture that fosters disease—particularly the plastic ones— and provide shelter for harmful insects and rodents, such as mice and voles. When left on for many years, these wraps can even strangle and kill trees.

Paper wraps or the more common white plastic guards should be applied in the autumn and removed immediately following the last frost. If you are using them primarily to prevent physical damage to the tree, you should replace them twice a year to ensure they are not compressing the tree's trunk or providing a home for pests or pathogens.

━● the real dirt *The usefulness of tree wraps and guards is a point of debate among researchers.* But those living in cold, harsh climates generally support the use of tree wraps as long as they are reserved for thin-barked, young trees and are applied and removed at the right times.

advice that's debatable Wrap evergreens in burlap for winter protection

Honestly, if you live in a cold, snowy climate and want to plant evergreens for winter interest, does it really make sense to wrap them up in burlap like mummies once the temperature drops? Wouldn't it be better to just plant deciduous shrubs and learn to love their bare branches? Apparently a lot of people do not think so, as even the quickest of Google searches turns up scads of gardening sites offering tips for using burlap to wrap evergreens.

Gardeners are also urged to try an even more unsightly approach in which they build ugly screens out of burlap to protect trees from sun and cold, drying winds. That's just what you want to look at all winter, right? Well, no, but we do have to admit that screens can help prevent evergreens from losing moisture by blocking the wind. Burlap wrapped around a tree, however, is not as effective because it can act as a big wick, sucking up what little water an evergreen has left in the winter, potentially creating even drier conditions.

WHAT HAPPENS IF YOU WRAP EVERGREENS IN BURLAP

Burlap is purportedly good at keeping evergreens safe from harsh sun and wind in the winter. It is also said to protect evergreens from street salt spray. Unfortunately, in practice, wrapping evergreens in burlap just is not that effective, and its wicking action may even create drier conditions. Burlap screens work much better.

If you do not wrap or provide a screen for evergreens during the winter, your landscape will certainly look better. But without screens, evergreens may be more susceptible to winter burn, depending on where you have planted them.

A BETTER WAY

Here are a few tips for minimizing winter injury to evergreens, whether you use burlap or not. Watering well through the autumn will go a long way toward protecting evergreens against winter burn. Site selection is also important. Evergreens such as arborvitae, yew, and hemlock should not be planted on the south side of buildings or in excessively windy or sunny spots.

If you really want to plant evergreens and are concerned about winter burn, don't wrap them but instead build an unsightly wall out of burlap for the first couple of winters. Once the plants are established, they should not need it anymore. The barrier should be constructed so it protects the evergreen from south and southwest sun, as well as wind, and it should be 3 or 4 feet from the tree itself. In other words, this could be one heck of a large barrier.

—• the real dirt *Wrapping trees in burlap is a hideous practice and a waste of time.* If you must use burlap, use it to build a wind-blocking wall.

advice that's debatable Remove twine, burlap, and wire before planting a balled-and-burlapped tree or shrub

Hold onto your shovels, folks. This is one of the most hotly debated bits of advice among tree researchers and others. Many things can go wrong when you are planting a tree, but leaving the twine in place that holds the top of the root ball together is a doozy. If that twine does not degrade properly (nylon twine never does, and sisal twine may or may not break down rapidly), you will end up strangling the tree and maybe even killing it.

In addition to removing the twine, many also recommended removing both the wire and the burlap that surround the root ball before planting. This is advised because tree roots can sometimes be constricted by the wire and/or burlap and get injured or otherwise deterred from growing as they should. But removing the burlap and wire can be problematic because, without those things, the root ball can fall apart, tearing young roots that the tree needs to take up water and nutrients. Still, we recommend removing as much of the wire and burlap as you can without the ball falling apart.

WHAT HAPPENS IF YOU REMOVE THE COVERING

The twine covering the top of a balled-and-burlapped tree can cause serious harm to a newly planted tree, so it absolutely must be removed. Though it has not yet been proven definitively, some researchers and landscapers believe that wire and burlap can harm tree roots, and they advise removing them at planting time.

If you plant a tree wrapped in twine, burlap, and wire and do nothing to free it before planting, it may be fine. Then again, it may not. Why take the chance?

A BETTER WAY

While you're working to free the tree from its confines, it is usually best to have a helper. Before planting, use scissors to cut the twine from the top of the root ball. You will need wire cutters to tackle the wire cage around the burlap. At the very least, you need to remove any wire that would be aboveground so that people won't trip on it. Removing wire lower down is good, but, again, don't remove so much that the ball of soil falls apart. Then use a box cutter or other sharp tool to cut away as much of the burlap covering the soil as possible. Basically, the idea is to remove as much wire and burlap as you can without having the ball of soil that the tree is in fall apart.

—● the real dirt *Though you will find plenty of people saying something contrary, it's really not a good idea to plant balled-and-burlapped trees without first freeing them from their bondage as best you can.*

advice that's just wrong: Top trees to keep them from getting too tall

Years ago, tree topping was considered a good way to keep trees from growing too large for a site. Also known as stubbing and heading, topping is the cutting back of a tree's leader and many of the upper, primary limbs, as well. When topping is done, there is no consideration given to where pruning cuts ought to be. Instead, everything is chopped back to a uniform height—as when city tree trimmers come around to cut trees back so limbs are kept out of the way of power lines. It's not a pretty sight.

We see topping done all the time by cities, tree service companies (even legitimate ones), and homeowners. Many people have heard that topping is a bad idea, but they don't hear the message consistently, or they don't know whether they should believe that advice or not. After all, their parents and grandparents topped trees. What's the problem?

WHAT HAPPENS IF YOU TOP TREES

Topping is a bad idea for a lot of reasons. When a mature tree is topped, a fair portion of the canopy is often removed. This significantly decreases the tree's ability to photosynthesize so it can make the food it needs to live. Worse still, poorly made cuts leave trees open to disease pathogens, pest infestation, and rot. And then there are the aesthetic problems.

In response to this extreme pruning, trees produce water sprouts immediately below the cuts. These fast-growing shoots not only look bad, they increase the possibility of wind and ice damage. Although these shoots may make the tree appear to grow just as lush, or maybe even more so than before, the tree will never regain its natural form. The new limbs will tend to have poor angles of attachment to the stem, making them very susceptible to breakage over time. This will be apparent in the autumn when the tree is bare. In short, topping is a disfiguring and completely unnecessary practice that may shorten the life of a tree or even kill it.

If you do not top trees, they will grow and mature as they were intended to and you will get to enjoy them in all their aesthetic beauty in your landscape.

A BETTER WAY

If you really need to reduce the height of a tree for some reason, a reduction cut is a better pruning method to use. Rather than making a straight cut across the top of the tree, remove long limbs at their point of origin (where they connect with a larger stem).

You could avoid the need to control the height of trees by simply choosing trees that are the right size for your site. When you do need to prune and shape trees around buildings or wires, hire a reputable company staffed by certified arborists who will prune correctly at the right time of year.

—● the real dirt *Do not top your trees.* It's harmful and unsightly. If you have a tree that is interfering with wires, blocking sun from a garden, or just giving you the willies because you are worried a limb might come down in a storm, you might be better off removing it and replacing it with a suitably sized tree than topping it. In all of these cases, consult with a certified arborist on the best course of action.

advice that's just wrong After pruning, dress tree wounds to inhibit decay and insect infestation

Drive around any neighborhood and you are likely to see at least a few trees with bright white paint or black tar covering pruning cuts Dressing tree wounds was long ago dismissed as a bad idea, yet homeowners still do it because some people are still promoting this strategy—often those selling the dressing products. There are even some eco-conscious companies offering natural healing products for this purpose. None of these products have been shown to be much better than doing nothing.

WHAT HAPPENS IF YOU DRESS TREE WOUNDS

Wound dressings are touted as necessary after pruning to protect trees from opportunistic insects and fungal pathogens that would take advantage of fresh pruning cuts. But why would plants need help with the process of healing? They have been handling it on their own for millions of years. To be clear, though, trees do not actually heal. When they are wounded, they isolate and compartmentalize the damage. Wound wood develops over the damaged area to protect it, and eventually the tree grows around the damage.

Dressing a tree wound with paint, tar, or some other product can actually seal in moisture, thus promoting decay, and prevent wound wood from forming.

A BETTER WAY

Do not use any type of dressing to cover tree wounds after pruning. The best way to avoid wound infections is to prune when pathogen problems are at their lowest, which is during the winter and early spring.

the real dirt *Wound dressings don't do what they are touted as doing and usually cause more harm than good.*

advice that's just wrong Prune trees when they are planted

This piece of advice seems to make sense, but it just doesn't work in practice. When a tree is planted it almost invariably has its roots compromised in some way. Perhaps they were cut off when the tree was harvested from the field. Maybe roots are circling the root ball because the tree was starting to get pot-bound. Or perhaps the roots of a pot-bound tree were sliced to try to get them to grow outward. The point is, there just are not as many healthy roots on a tree that is about to be transplanted as there are on a tree that has never been transplanted. Some garden gurus reason that because roots supply the tree with water and leaves use the most water, it makes sense to prune off some of the tree's canopy. This, they argue, keeps the limited amount of water the roots are able to supply from being wasted.

The problem with this theory is that trees do a fine job of compensating for their lost roots without our interference. Unlike perennials, trees are usually dormant when they are transplanted or their roots are compromised, so when their buds break they break with a reduced root system in place. A transplanted tree may drop some of its leaves or a certain percentage of leaves will not flush out at the beginning of the season. Leaves may also be smaller if a tree's root system has been compromised. Pruning limbs off of the tree does not really help it cope with root loss, and it may even create problems because of the number of open wounds that could become infected with a potentially deadly disease.

WHAT HAPPENS IF YOU PRUNE TREES AT PLANTING

It is possible to have problem limbs that need to be removed at planting time. Limbs can get damaged when trees are moved around. A limb or two might even be dead or growing at a bad angle that makes it rub up against or hug another limb too tightly. In these cases, pruning is a good idea. Opting not to prune a tree at planting time is not going to have much effect on whether the tree lives or dies.

A BETTER WAY

If you do decide to prune, remove only problem limbs. Pruning for shape can be done later, once the tree has become fully established in its new site. Be sure to make a smooth cut a little bit beyond where the branch intersects the tree's trunk. This area is called the branch collar, and it's best not to cut it if possible.

━▶ **the real dirt** *We do a lot of tree planting, and we rarely prune off any limbs unless they are faulty.* When roots are compromised at planting time, the tree usually reduces the size of its leaves. In fact, we have seen maple trees that had their roots extremely compromised (less than 10 percent of their original root system was left intact) do just fine after planting. The only noticeable issue was that the leaves of these maples were only about a third the size of those of other maples. These small-leafed trees looked like Japanese maples, but they recovered after a couple of seasons.

advice that's just wrong Beat a tree to get it to flower

When a plant is under severe stress, its natural response is to try to produce offspring. In other words, the plant uses all of its energy reserves to set flowers that will later become seeds. One way gardeners can capitalize on this tendency is to put trees under stress to force them to flower. The problem with this plan is that if you are doing enough damage to get a tree to flower, you are probably also harming it to the point of damaging its health. Still, some garden experts suggest beating a tree's trunk with a baseball bat, broom handle, or some other such thing to get it to bloom. This does work, but come on!

WHAT HAPPENS IF YOU BEAT A TREE

When you beat a tree with a bat you damage its vascular tissue, the part of the plant that transfers water and nutrients up the stem and the products of photosynthesis down the stem. If you really don't like your tree and you plan on removing it, but maybe want to keep it for one more season, pull out the old baseball bat and have at it. If you don't beat your trees, you may have fewer blooms, but your trees will be much healthier.

A BETTER WAY

Trees will bloom in their own time. For some trees, such as apples, it may be 15 years before a tree flowers, but it will eventually flower. Some trees, like oaks, only flower well every other year. Beating them to try to make them flower more often is not a good idea.

—● **the real dirt** *Don't ever beat your trees.* You are most certainly doing more harm than good, and your neighbors will think you are scary.

advice that's just wrong
Make pruning cuts flush with the trunk

A long-standing misconception about pruning trees is that they should be pruned so the cut is flush with the trunk. Aptly called the flush cut, this technique is considered one of the worst ways to prune a tree. When you make a flush cut, you expose more of the tree's vascular tissue than with a proper pruning cut, increasing the chance for disease and slowing the healing process. A proper pruning cut is made right above the thickening of the branch, called the branch collar, which occurs at the intersection with its limb of origin.

WHAT HAPPENS IF YOU MAKE FLUSH PRUNING CUTS

Pruning by making flush cuts increase the plant's risk of disease and slows the formation of wound wood around the cut. You might consider a flush cut if you cannot find a clear branch collar. However, it's actually better to cut the limb a little bit further down its length than to cut it too close to its limb of origin.

A BETTER WAY

To make a proper cut, use pruners or a saw to cut the limb just above the branch collar. It is always safer to cut a little bit further down the branch than to make a flush cut.

the real dirt *Let the branch collar be your guide when pruning a tree branch.*

....

advice that's just wrong Shear the tops of deciduous shrubs to control their height

Shrubs are a beautiful addition to the garden landscape, and it's so easy to completely ruin their natural shape. Let's say your lilacs are getting really tall, so you go out with an electric hedge trimmer or shears and hack 3 or 4 feet off the top. Sure, it's quick and easy, and it definitely cuts those unruly shrubs down to size. But then what happens? New shoots grow like crazy where you make those cuts. Just like tree topping, this technique stimulates a lot of new growth directly under the cuts. Unfortunately, this new growth is not the kind you want.

Shearing is not pruning. Although it's fine to shear some types of formal hedges, such as boxwood *(Buxus sempervirens)* and arborvitae *(Thuja occidentalis)*, it should never be used as a strategy for shaping deciduous shrubs, such as lilac *(Syringa)* and *Forsythia*. All of that out-of-control growth that follows shearing destroys the natural form of deciduous shrubs. If shearing is done at the wrong time of year, for example, at the beginning of the season for shrubs that bloom on old wood, it will likely result in much less flowering, too. Instead of shearing, deciduous shrubs should be pruned with thinning and renewing in mind. This will not do much to change the height of plants, but it will keep them healthy and looking their best.

WHAT HAPPENS IF YOU SHEAR THE TOPS OF SHRUBS

We understand why people want to control the height of tall shrubs. But when you shear shrubs, you create a lot of unsightly growth and leave stubs all over the place that will not heal well like proper pruning cuts do. This can lead to disease and insect problems pretty quickly.

If you don't shear your shrubs they may be taller than you would like, but they will be healthy, much more likely to flower well, and less prone to disease and pest problems.

A BETTER WAY

When pruning deciduous shrubs, it's best to use a hand pruner, and you may need lopper shears for bigger branches. Prune spring-flowering shrubs immediately after they have bloomed because many form the following year's flower buds on the current season's growth. If you wait until later in the summer to prune, you will be cutting off the following spring's flowers. In general, however, late winter and very early spring is a good time to prune to avoid possible disease issues.

The two main pruning goals when dealing with deciduous shrubs are thinning and renewal. Remove old, damaged, or overcrowded branches by cutting them off at the base of the plant close to the ground. If you have a very old shrub that is badly in need or rejuvenation, it's safe to remove about one-third to one-half of the oldest stems by cutting them off at the base. You can remove another third (or the rest) the following year. Cutting off all the old branches at one time is too hard on a plant, particularly one that is been neglected for a long time.

A better option would be to choose shrubs carefully from the start, being mindful of their mature size before you bring them home. If you won't be thrilled about having a 10-foot-tall lilac hedge along your sidewalk, then go with something shorter, such as ninebark *(Physocarpus opulifolius)* or *Viburnum*, instead.

━● **the real dirt** *Deciduous shrubs are set to grow to particular height ranges.* You can control height a little bit by pruning the correct way, but not much. If you resort to quick-and-dirty shearing to cut deciduous shrubs down to size, you are going to ruin their natural form. In the end, you'll create more work for yourself as you try to keep them alive and looking even halfway decent.

7

VEGETABLES
and FRUIT

 # In This Chapter

Good Advice

∘ Do not prune fruit trees in the spring and summer

∘ Incorporate eggshells into the soil to control blossom-end rot in tomatoes

∘ Rotate vegetable crops from the same family regularly

∘ Thin seedlings for healthy plant growth

∘ Weed regularly to keep garden plants healthy

Advice That's Debatable

∘ Plant hybrid rather than nonhybrid vegetable seeds

∘ Do not use treated lumber around edible plants

∘ Plant vegetables in rows

∘ Plant vegetable crops during the correct phase of the moon

∘ Use companion planting to help control pests

Advice That's Just Wrong

∘ Don't plant vegetables in containers

∘ Only plant vegetables in full sun

There's really nothing like the feeling of growing your own food. Walking out the door to collect tomatoes for a pasta, cucumbers for a salad, or raspberries for a pie is a joy not to be missed in this life, even if you do eventually decide you'd rather not fuss with some things (like asparagus), figuring it's easier to just buy them.

Maybe it's our long history of growing vegetables out of necessity that makes this type of gardening more prone to disagreement than others. Or maybe it's because growing food is so close to our hearts. Gardeners who tend vegetables are often carrying on family traditions, growing things the same way their parents did, and their parents before them, and so on. Plenty of heirloom varieties can be traced back to family members who brought seeds to America from their homelands by sewing them into skirt hems and hat brims for safe passage through Ellis Island.

There is no arguing with memory and heart. Vegetable gardeners exist in a world of fact and folklore, their crops grown in traditional ways that are not always possible to analyze scientifically. Still, we have done our best to sift through some of the advice we hear regularly. No offense to grandma.

good advice Do not prune fruit trees in the spring and summer

Pruning not only helps control the size and shape of fruit trees so they can develop the framework they need to support fruit production, it improves the quality of fruit and boosts yield. Pruning also ensures good air circulation and makes it possible for sunlight to penetrate all parts of a tree's canopy. Like any type of tree, fruit trees should be pruned at a certain time of the year. There are a lot of opinions on when to prune, but those who advise against pruning in the spring and summer have it right.

There is one exception to this bit of pruning advice. If you see suckers at the base of a fruit tree or below the graft on a tree, prune those right away. Unless is it especially hot and humid and you might encourage disease problems by pruning, do not wait until suckers grow long and gangly. Cut them off as soon as you see them.

WHAT HAPPENS IF YOU AVOID PRUNING FRUIT TREES IN SPRING AND SUMMER

Gardeners who do not prune their fruit trees in the spring and summer will greatly reduce the likelihood of diseases like fireblight, which fruit trees are more susceptible to in hot, humid conditions. By waiting until at least late autumn to prune, you will also avoid encouraging new growth that will not have time to harden off before winter.

While it's true you can prune anytime you like, choosing the right time to do the bulk of your pruning can mean the difference between a healthy tree and one that is suffering. It is fine, though, to do small things, like remove a single limb that is diseased, injured, or dead.

HOW TO DO IT

A good time to prune fruit trees is in late autumn and early winter. But we would advise pruning later into the winter, when you are past the coldest weather but before bud break, because wounds will not seal up quickly while the trees are dormant. The longer those cuts stay open, the more likely it is that the tree will be damaged by dehydration. Another benefit to winter pruning is that there are no leaves to worry about, so it is much easier to see the shape of the tree. Dead branches and crossing branches will also be easier to spot.

When pruning, remember to do so in a way that allows you to reach the fruit as it matures. Sterilize your tools in a solution of 1 part bleach to 10 parts water between cuts, or at least between trees, to avoid spreading disease. Because sealing tree wounds has not proven to be an effective way to prevent disease and may even slow wound healing, we advise against doing that.

—● **the real dirt** *Except in the case of suckers or small, injured branches, it's best to wait until the late winter to prune fruit trees because you significantly reduce the chance of disease.*

good advice Incorporate eggshells into the soil to control blossom-end rot in tomatoes

Few things make gardeners who grow tomatoes as crazed as when they discover blossom-end rot. All that fussing, watering, fertilizing, and waiting and there they are: black, mushy sunken spots on the backsides (blossom ends) of some tomatoes. The problem is caused by a calcium deficiency. Tomatoes, as well as peppers and eggplants, need to take up calcium from the soil to produce fruit.

Lots of things can prevent a plant from taking up enough calcium, most notably insufficient calcium in the soil or growing media. But even if enough calcium is available, other things can disrupt proper calcium uptake, such as dry conditions or the overuse of nitrogen fertilizer. Much of the time, though, blossom-end rot is due to inconsistent watering, so soil moisture is in a constant state of flux. It is water that carries calcium throughout the tomato plant.

Without adequate calcium, tissue breaks down and those awful black lesions start to form. Adding crushed eggshells to the soil helps stave off blossom-end rot because the eggshells provide an extra shot of calcium to the plant. Vegans and others who don't have eggs around the house can add gypsum (calcium sulfate) to the soil, instead of eggshells.

WHAT HAPPENS IF YOU INCORPORATE EGGSHELLS

Although you can safely eat tomatoes that have blossom-end rot by cutting off the blackened part, it's not very appetizing, to say the least. You can avoid the whole mushy mess in the first place by watering regularly and adding some form of calcium to the soil before you plant the tomatoes.

If you forget to add any calcium, and even if you're doing a good job of watering, blossom-end rot can strike. When it does, we always add a few crushed eggshells to the soil and the problem goes away. So don't panic if you see blossom-end rot on the first tomatoes of the season. That is not uncommon. With proper watering and additional calcium, the rest of your tomatoes will likely be delicious and free from the black stuff.

HOW TO DO IT

There is no standard recommendation for the number of eggshells that should be added to the soil to reduce the chances of blossom-end rot. While some people say two eggshells every couple of weeks, others say five and even ten. We recommend adding about four crushed eggshells to the soil per plant at planting time. This is particularly important for container-grown tomatoes, but can be a good idea in the garden, too. Adding a little compost to the potting media when planting container-grown tomatoes is also a good idea because it will provide nutrients as it breaks down. Of course, consistent watering is also a key to keeping this disease away.

—● the real dirt *Crushed eggshells add calcium to soil or container media and help prevent the onset of blossom-end rot.* Proper watering is a fundamental part of avoiding this problem, as well. The eggshells will not do it by themselves.

good advice Rotate vegetable crops from the same family regularly

People have been rotating crops for as long as they have been planting fields. Before pesticides and herbicides, crop rotation was the method farmers used to control soil-borne diseases and manage insects. This practice is still quite useful today, though it can be difficult for home gardeners who don't have a lot of land to work with.

The problem with using the same site for the same crops year after year is that pathogens specific to the plants grown there build up over time. Insects that are partial to those plants will also thrive and return to the spot. Eventually, you will have serious problems on your hands. By rotating crops every other year or so, you break this cycle, causing pathogens to die and the insects to move on.

WHAT HAPPENS IF YOU ROTATE VEGETABLE CROPS

Crop rotation is no miracle cure. It will not completely prevent disease or insect pests from harming vegetables, but it will help. It is also a good way to keep soil healthy and nutrient rich. Crops have different nutritional needs, so they take different things from the soil. Leaving one crop in place for a long time may deplete that area of specific elements. You could correct these deficiencies with fertilizer, but you would need a soil test to know what to add.

Another good reason to rotate crops is that plant roots penetrate the soil in different ways and at different depths, depending on the crop. This affects the makeup of the soil, as well as the activity of the microbes that live there. By moving crops to different parts of your garden every year, you are essentially spreading around the benefits (and drawbacks) each plant has to offer.

Crops will be more likely to suffer from the effects of the diseases and insects that afflict them if they are not rotated. When tomatoes stay in the same spot year after year, for example, you are more likely to see diseases such as early and late blight and *Verticillium* wilt appear. Cucumbers that are not moved are more likely to suffer from cucumber mosaic virus, which is most often transmitted by aphids feeding on plants in the area and transferring the virus from plant to plant.

HOW TO DO IT

Before planning how you will rotate what you grow, start by learning which plants belong to each family. For example, tomatoes, peppers, potatoes, and eggplant are all part of the Solanaceae family, whereas the Cucurbitaceae family includes cucumbers, melons, squash, pumpkins, and gourds. Legumes like peas and beans are part of the Fabaceae family, and the Brassicaceae family counts Brussels sprouts, kale, and broccoli among its members.

The ideal way to rotate crops would be to divide them up by family, then make a plan for rotating them in your garden every season while keeping in mind that plants in the same family should not occupy the same spot again for about four years. To do this, you could move clockwise or in some other pattern. It does not really matter as long as you are keeping plants on the move, particularly the plants in the Solanaceae, Cucurbitaceae, and Brassicaceae families.

If you don't have the space to do this much moving, you could rotate some plants like tomatoes and peppers into large pots in a sunny location. Or do your best to find ways to intersperse vegetables into different parts of perennial gardens each season. Without driving yourself nutty, just try to rotate what you are growing. If you cannot manage this every year, do it every other year.

—• **the real dirt** *Vegetables in the same family are less likely to get diseases if they are rotated.* When Meleah was new to vegetable gardening, she did not take the recommendation seriously. Within three years her tomatoes, cucumbers, and squash were all sick with diseases that might have been prevented had she moved plants around.

good advice Thin seedlings for healthy plant growth

When it comes time to thin vegetable seedlings, the saying "Kill your darlings" comes to mind. Attributed to both William Faulkner and Mark Twain, the morbid directive is meant for writers who cannot easily part with anything they have created. But the meaning applies here, too. Get rid of those tiny seedlings that seem precious but are really just mucking things up.

This is much easier said than done for many gardeners. It's hard to just kill some plants while letting others live. We would much rather keep every seedling. But this usually is not a viable option, so thinning must be done to ensure the healthy growth and development of plants. Otherwise, there are just too many plants in one spot competing for the same water, nutrients, and space.

WHAT HAPPENS IF YOU DON'T THIN SEEDLINGS

Whether seeds are sown in trays or outdoors, or they germinate after self-sowing, it's not uncommon to have several seeds in one planting hole. This is a problem because every plant has its own space needs. Overcrowding taxes the long-term health of plants. Not only will there be too much competition for water and nutrients, plants will be so close together they will not be able to get the air circulation and sunlight they need. Diseases can take hold more easily under these conditions, and when they do, they spread quickly.

HOW TO DO IT

When seedlings emerge from the soil, the first set of leaves are the cotyledons. These leaves look the same on most seedlings, regardless of the species. Soon, though, the cotyledons wither and drop off and the seedlings first true leaves appear. Once you see these true leaves, it's time to thin the seedlings.

If you want lots of plants, you can gently separate each seedling from the group in the pot and replant each one in a separate tray or pot for later transplanting into the garden. Work quickly to avoid having the seedlings dry out. If you are thinning seedlings in the garden, you can simply separate them from the group and transplant them directly into the soil elsewhere. If you want to follow the particular spacing requirements, check the back of the seed packet for that information.

If you are not keen on having a million new plants, there's an easier way to thin seedlings. From each group choose one that looks good and strong, then use scissors to cut the rest off at the soil level. This works in the garden and for seeds sown indoors. By using this method rather than plucking out the unwanted seedlings, you are not disturbing the root systems of developing plants.

If you just don't have the heart to thin seedlings in pots or in the garden, you can carefully place one seed in each hole at the right spacing to start with. The problem, though, is that the germination rate of seeds is usually far from 100 percent, so you may wind up with fewer plants than you'd hoped for.

➥ the real dirt *Thinning seedlings always pains us a bit because we do not like to kill things.* But we know it must be done. If you let too many seedlings grow in one spot, the stress of growing in such cramped quarters will adversely affect the mature plants.

good advice Weed regularly to keep garden plants healthy

Some people call weeds "the right plant growing in the wrong place." That is accurate, but it sure feels unduly kind when you've been squatting for hours in the heat trying to rid your garden beds of them. Some weeds that were brought here from other countries because of their pretty flowers, such as creeping bellflower *(Campanula rapunculoides)*, are beloved by some gardeners, even though they can be invasive. Others, such as broadleaf plantain *(Plantago major)*, are unquestionably just weeds. And weeds—no matter how attractive—cause problems in the garden because they compete with desirable plants for water, air, light, and nutrients.

WHAT HAPPENS IF YOU DON'T WEED REGULARLY

Taking a live-and-let-live attitude with weeds will save you backaches, but it will eventually cost you your garden. They may move rapidly to take over the place, or they may do it slowly, but all weeds deprive garden plants of what they need to live. Some, such as bindweed *(Convolvulus arvensis)*, kill the plants in your garden in an even less subtle way. Aptly named, bindweed twines tightly around plants and literally smothers them.

HOW TO DO IT

Taking proactive action to keep weeds from getting too firm a hold in your garden in the first place is the best thing to do. Mulch can be very helpful in controlling weeds. When it is applied thickly (3 to 4 inches is good), weed seeds cannot germinate nearly as fast as they can on open ground. Of course, if you already have weeds in the area you are mulching, you will have to pull those first or they will just come up through the mulch after a while. People throw mulch down over weeds all the time hoping it will smother them, but it does not. Trust us on this.

No matter what you do, you are going to have some weeds. The best way to deal with annual and perennial weeds is to use a hoe to separate their tops from their roots early in life—before they get established enough to grow back or spread by seed, suckers, or underground roots. If you wait too long, you will have to pull them. Dig down deep to get all of the roots, particularly if the weed has a taproot, such as dandelions *(Taraxacum officinale)*, or it will just grow back. While it will be tempting to reach for a pre-emergent weed killer, such as trifluralin, this really is not an option in a garden where you are encouraging young plants to grow.

—▶ the real dirt *Weeding is just part of gardening. It's not the fun part, but it has to be done.* You don't have to pull every weed you see, but you need to get rid of most of them. Before you plot out your garden, be honest with yourself. If you hate the idea of weeding, don't dig up enormous parts of your yard to create garden beds. Plant what you know you can reasonably take care of. You will enjoy your gardens much more, instead of viewing them as just another chore.

advice that's debatable Plant hybrid rather than nonhybrid vegetable seeds

It's common to see the word **hybrid** *on vegetable seed packets.* Hybrid plants are a cross between cultivars (cultivated varieties) or strains of the same species that have different traits. (In some cases, hybrids may even be crosses between different species.) While hybridization is often deliberate, it can happen by accident thanks to the movement of pollinators or the wind.

When plants are deliberately hybridized through pollination, the goal is to improve one trait or another. Sometimes those improvements are good for gardeners. Other times they benefit commercial producers who might be willing to sacrifice taste, for example, in order to have tomatoes with tougher skins so they travel better.

WHAT HAPPENS IF YOU PLANT HYBRID VEGETABLES

With vegetables, hybridization can mean increased disease resistance, higher yields, earlier maturity, or better flavor. The upside of hybridization is that the good genetic traits of two parents are passed along, so you get plants that are more uniform, more productive, more vigorous, and better able to handle heat and cold.

The benefits of hybrid plants are hotly debated, however. Some argue they are a vast improvement over heirloom varieties grown by our ancestors. Others say hybridization is the reason why these old favorites, which may taste better and produce for longer periods, are getting more difficult to find. If you don't buy and plant hybrid seeds, you will not enjoy the benefits they offer. But you will save money because the effort it takes to pollinate plants to create hybrids is reflected in their price.

But price is not the greatest strike against hybrid seeds for many gardeners. What gardeners often don't like is that they cannot collect the seeds of hybrid plants and save them from year to year. This is because hybrid seeds do not grow true to type when planted, meaning the offspring will not be the same as the parents. So if you want the same plant from year to year, you have to buy the hybrid seeds again and again— or make your own hybrids, which is a tricky proposition for most gardeners.

A BETTER WAY

It's up to gardeners to decide what types of plants and seeds to buy. There are reasons for buying hybrids, chief among them being their resistance to certain diseases and harsh environmental conditions. One reason for planting heirloom varieties is that they frequently have better flavor. Also, gardeners who buy heirlooms help preserve the kind of plant diversity that might one day be vital to our future food supplies. While hybridized plants can be bought almost anywhere, heirlooms are harder to find. Two good sources are Seed Savers Exchange (http://www.seedsavers.org) and Seeds of Change (http://www.seedsofchange.com).

the real dirt *When it comes to vegetables, we usually buy heirloom seeds and plants whenever we can.* We have had both good and bad luck with them. Unlike hybrids, heirlooms can be unpredictable, less disease resistant, and more susceptible to temperature extremes. Still, the flavor of heirloom vegetables is usually much better than that of hybrids. Using heirlooms can also be satisfying in that it allows you to be part of a community that trades and plants seeds that have been selected and handed down by generations of home gardeners.

advice that's debatable Do not use treated lumber around edible plants

Pressure-treated lumber has been used as garden edging and raised beds for decades. But is it safe for use around edibles? In December 2003, the U.S. Environmental Protection Agency banned wood treated with chromated copper arsenate (CCA) for residential use due to concerns about the possibility of arsenic leaching into the soil in amounts that could be harmful.

Taking its place was another type of treated wood dubbed ACQ for the alkaline copper quaternary compounds it is treated with. The lumber contains no arsenic that could leach into the soil, but studies have shown that some copper leaching takes place. Copper is much less toxic than arsenic, but whether ACQ belongs near food crops is continually debated.

WHAT HAPPENS IF YOU USE TREATED LUMBER

The big advantage to treated lumber is that it breaks down very slowly so it lasts a long time. Even as it was being pulled from the residential market, there were plenty of scientists around insisting that the arsenic contained in CCA-treated wood was not present in the soil in harmful amounts in most cases. If anything, they said, the only area to be concerned about was the foot or so closest to the wood. With ACQ, the only cautionary advice consistently offered by scientists and the Environmental Protection Agency is that it should not be used near ponds and streams because copper is toxic to aquatic life.

There are alternatives to pressure-treated lumber, such as cedar. But cedar is more expensive and will break down more quickly than treated lumber. Recycled plastic can also be used, but there is debate about the safety of this material because some types of plastic leach bisphenol A (BPA). Studies have shown that the rate plastic leaches BPA increases rapidly when the plastic is heated, as it would likely be in the summer sun. Although the effects of BPA are still being studied, it has been established that this chemical is an endocrine disruptor, that is, it mimics the effect of hormones in the body. BPA does make it into the soil, but it breaks down rapidly. Likewise, the BPA that may be taken up into plants is broken down rapidly, so it is highly unlikely that it is transferred into vegetables or fruits. Still, some people feel that is it's better to be safe than sorry.

A BETTER WAY

Cedar, natural stone, rocks, or landscape block can also be used to create raised beds. To some, these are the only reasonable choice when growing vegetables and other edible plants.

If you have concerns about the safety of ACQ, but you still want to use it for a raised bed, some recommend lining the sides of the bed with heavy plastic to reduce the risk of chemicals leaching into the soil. But, again, you are bringing plastic into the picture and it may leach unwelcome chemicals, such as BPA.

—● the real dirt *There is nothing to indicate that eating vegetables grown near lumber treated after 2003 will be toxic to people in any way.* However, some are concerned about this lumber and would rather be safe than sorry. If you are one of these people, consider using stone, landscape blocks, or cedar.

advice that's debatable
Plant vegetables in rows

Gardeners use several styles successfully for growing vegetables. Although planting in traditional, long rows is just one of them, it is the most recommended planting method. This likely has more to do with tradition than success rate because wide rows (rather than long, thin ones) and square-foot-gardening are also good options.

Whatever style of planting you choose, spacing between plants should always be determined by the plant's mature size, which can be found on the plant's label or the back of the seed packet. If you're gardening in traditional rows, it's a good idea to create rows that are around 20 to 22 inches wide for trellised and caged plants, like tomatoes and cucumbers. Even more width is needed for vining crops like squash. In larger gardens, pathways between rows should always be at least 16 inches wide to allow a wheelbarrow to get through without running over plants.

Followers of the square-foot-gardening method made popular by Mel Bartholomew, author of *Square Foot Gardening*, are interested in trying to grow as many vegetables as possible in a small space. This gardening strategy is also popular because it is based on the use of raised beds, which are often easier to tend, particularly for the elderly or people with disabilities. These intensely planted raised beds can also be a lifesaver for gardeners dealing with poor soil.

In square-foot gardening, everything is planted on a grid: herbs might be in one square, for example, while tomatoes are in another. Plants that would normally take up a lot of space on the ground, such as cucumbers and squash, are grown vertically. Not only can you grow a lot of vegetables and herbs in a small space, you use fewer resources when tending them.

Wide rows are also an option for vegetable gardeners, particularly those with smaller gardens. This gardening style is essentially just what it sounds like, instead of long, thin rows gardeners plant several rows of the same thing together in what could best be described as a block. Plant spacing depends on the type of crop, and directions on plant labels and seed packets should always be followed to ensure plants aren't too close together. Like square-foot gardening, this method saves space and resources, as well as time. Limiting beds to a width of 4 feet is a good idea so you can reach all the way across.

WHAT HAPPENS IF YOU PLANT VEGETABLES IN ROWS

Traditional, straight rows do have advantages. You can walk between rows, making harvesting and care easy. Insect control is also a lot simpler because plants are not crowded and pests can usually be spotted quickly. If you have to use chemicals to control insects, it's easy to do that with rows because you're unlikely to spray neighboring plants.

If you don't plant in rows you will be able to produce more vegetables and herbs in less space, which is a plus. But be mindful of the fact that plants prefer different growing conditions. Corn and tomatoes, for example, have different habits and different needs. Before choosing a style of vegetable gardening, consider your space, the plants you want to grow, and your ability and/or willingness to weed, water, and construct raised beds.

A BETTER WAY

To plant in slender, straight rows, stretch some rope tightly between two stakes at the end of each row. Make furrows for the seeds as you plant. Be generous with the seed because all of them will not germinate, but don't plant too many in one hole.

If you like the idea of easy-to-tend rows but want to make more of your space, try planting in wide rows or build some raised beds and try square-foot gardening.

—• the real dirt *Having grown vegetables in tradition-al rows and in wide rows or blocks, Meleah prefers the latter method if only for the fact that she produces more food with less space.* It's a bit more difficult to tend the wide rows because plants are closer together, but it's worth it.

....

advice that's debatable Plant vegetable crops during the correct phase of the moon

Some gardening practices are based on science, whereas oth-ers are more folklore than actual fact. Both have their merits, and even the most experienced gardeners can attest to the fact that some of the wackiest-sounding advice can work. This brings us to the strategy of gardening according to the phases of the moon.

Proponents of lunar gardening say it is not just the tides that are affect-ed by the moon's phases. Crops are under the moon's influence, too. Yes, it sounds a little hokey, but when searching for gardeners to talk about their experiences with lunar gardening, we were amazed by the amount of email that flooded our inbox.

"People think I'm crazy, but my grandmother always maintained that you should plant root crops during the dark of the moon in May and the rest could be planted safely after the full moon," one gardener wrote. Another longtime gardener chimed in, "Here I go again with the good old days. In the forties, we planted with the moon phases. Our grandparents and neighbors were all pioneers from the Old Country. It's been since poo-pooed and science has dominated the planting cycles . . . but I remember my mother saying that we had to hurry to plant certain things in the new moon." You can see why this is so hotly debated. Tradition is an important part of many people's lives.

WHAT HAPPENS IF YOU CONSIDER THE LUNAR PHASE

If you want vegetable and fruit crops to thrive, believers in gardening by the cycles of the moon say they should be planted during the right phase. Many believe harvesting is also best done according to the moon's cycles. Lots of gardeners grow edible crops while paying no mind to the moon, so you can, too. But what if this practice does offer some value? Aren't you even a little curious?

A BETTER WAY

There are many books on gardening by the phases of the moon. Some are easier to read and follow than others, largely because some get bogged down in a lot of talk about the relationship between astrology and the moon. Frankly, that's just a little too esoteric for us. Honestly, the principles for gardening by the lunar calendar are pretty straight-forward, so you may not even need a book. Instead, seek out helpful websites that offer current lunar calendars to help guide your plant-ing. We found the Gardening by the Moon website informative and helpful (http://www.gardeningbythemoon.com).

Here are the basics of what you need to know. Plants with edible parts aboveground, particularly those with seeds outside their fruit (such as spinach, celery, broccoli, cabbage, and lettuce), should be planted during the new moon. This is the first of the moon's four cycles and the start of the moon's waxing phase (when it appears to grow larger as it becomes full).

The moon's gravitational pull lessens a little during the second quarter of its waxing phase, but the strong moonlight supposedly helps promote leaf growth and helps with seed germination. This is a good time for planting crops with edible parts aboveground that form seeds inside the fruit, such as squash, peppers, peas, melons, and beans.

Plants with edible parts underground (root crops such as potatoes, beets, onions and carrots) should be planted during the third quarter of the cycle when the moon is waning or shrinking into the shape of the letter C. There is less moonlight after the full moon, but the gravitational pull remains high enough to supposedly keep the amount of moisture in the soil high.

Planting should not be done, if possible, during the fourth quarter, as the moon completes its waxing phase. With very little moonlight and less moisture in the soil, this time is best spent weeding, harvesting, and transplanting. Weeds are said to come out the ground much more easily during a full moon.

—● the real dirt *We have never tried it, so we have no first-hand knowledge of whether gardening by the phases of the moon works.* But we know a lot of gardeners who use this method because their ancestors did, and they believe it works. So we don't see any reason not to try it. Just don't start telling us that your beans are Virgos.

advice that's debatable Use companion planting to help control pests

So much is said about the benefits of companion planting, it's hard not to dismiss the idea as some sort of strange miracle cure for everything. Although there is not much science to back up the idea, many gardeners say they have had positive results when grouping certain types of plants together. The best-known work on the subject is probably Louise Riotte's book *Carrots Love Tomatoes and Roses Love Garlic: Secrets of Companion Planting for Successful Gardening.*

Some of the purported benefits of companion planting are deterring insect pests, providing food for beneficial insects, and improving flavor. This last one seems a bit sketchy, but we are including it because we know gardeners who swear it's true. For example, they say to improve the flavor of carrots, plant them next to chives.

As far as science goes, research does support using companion planting for pest control, at least to a certain extent. Studies have shown that when some types of plants are grown together, pests have a more difficult time moving from one plant to another or finding a host plant to feed on because nonhost plants are blocking the way or confusing the pest. Researchers have also found that putting certain plants next to others can attract natural predators that help control pests. Geranium, when planted next to cabbage, is said to control cabbage worm. Basil is believed to control aphids, mites, asparagus beetles, and tomato hornworm when planted with tomatoes or asparagus. Planting chives or garlic with carrots and tomatoes is supposed to repel aphids, mites, and weevils. Beware, though, some combinations can do more harm than good. For example, chives are said to stunt the growth of beans.

WHAT HAPPENS IF YOU USE COMPANION PLANTING

If you want to grow healthy vegetable crops without using chemicals, companion planting might be helpful. Gardeners and farmers have been using companion plants for a very long time and reporting success, and there is some science to support this concept. The strategy is simple, straightforward, and easy to implement for the home gardener. Plant basil and tomatoes together to enhance the growth and flavor of both crops. Or plant tomatoes with asparagus to repel asparagus beetle. While neither of the effects listed above has been scientifically proven, it's not a stretch to believe that they may be at least somewhat effective.

If you are the kind of gardener who likes neat, orderly rows of the same plant, you don't need to try companion planting. It's fine to keep on doing what you are doing if it's working. You may have to use more chemicals to control pests if you go this way, though.

A BETTER WAY

Start by doing your homework, because some plant combinations work and some do not. Talk to gardeners who use companion planting and read up on the subject. You can also try your own plant combinations, and we encourage this. As you will discover, there are many ways to do it. The object is not necessarily to have the two different plants right on top of each other, but rather to alternate the two plants within a row. Plants also can be scattered randomly across a garden to good effect. However, this may make organizing your harvest more difficult.

—● the real dirt *We know many gardeners who use companion planting year after year and report great results.* While the science does not necessarily support all of the combinations that gardeners try, it does support the idea that planting different plants together can help reduce pest problems. We see no reason not to try this.

advice that's just wrong Don't plant vegetables in containers

We hear this quite a bit and, thankfully, this notion is being debunked as more and more people are growing some of their own food, even in small spaces. Vegetables can definitely be grown successfully in containers, as long as you are mindful of a few basics, namely, sunlight, water, fertilizer, and pot size. As long as the site has enough sun and you choose the right pot and water and fertilize well, you've got it made.

WHAT HAPPENS IF YOU PLANT VEGETABLES IN CONTAINERS

Gardeners grow vegetables in containers for a lot of reasons. Maybe you don't have space in the garden, or you want to use that space for something else. Maybe you need to rotate crops and you cannot unless you move some things into containers. Or maybe you've got no garden at all and just want to grow a few edibles on a balcony or doorstep. Whatever the reason, it's easier than some make it out to be.

A BETTER WAY

Pot size is everything. Favorite vegetables like most tomatoes, beets, sweetpotatoes, peas, cucumbers, broccoli, carrots, and squash need large containers (at least 20 inches wide and 16 inches deep) that offer room for roots to spread out. Peppers, eggplant, and some compact tomatoes can be grown in pots that are about 15 inches wide and 10 to 12 inches deep. You can grow lettuce, herbs, arugula, cress, and compact peppers such as habaneros in containers that are 10 inches wide and 10 to 12 inches deep. Always choose containers with good drainage holes at the bottom or plants could get waterlogged and literally drown in the water you give them. And do keep in mind that vegetables have different needs for light. Even in containers, plants such as tomatoes need full sun, so make sure the container is placed in a spot where the plants will get sufficient light.

The biggest challenge to growing vegetables in pots is making sure they get enough water and nutrients. Containers dry out faster than garden soil, so it's important to water daily, sometimes twice a day in very hot weather. Because water washes away nutrients quickly, you will need to fertilize regularly, too. Granulated slow-release fertilizers are a good choice for containers. They feed plants over a long period of time, unlike the quick-release kinds that mix with water and offer more instantaneous food.

Organic fertilizers will likely require more applications, whereas slow-release, synthetic fertilizers will last for several months. Fish emulsion, an organic option, can also be used twice a month for added nutrients, if you like. Be sure to start with good potting soil, because garden soil is too heavy for pots. If you are growing tomatoes, calcium is a must, so be sure to add eggshells or gypsum.

—● the real dirt *If you want to grow vegetables, containers are as good a place as any as long as you are mindful of pot size, sunlight, and water, and fertilizer needs.*

advice that's just wrong Only plant vegetables in full sun

We hear people lamenting the fact that they cannot grow vegetables because they do not have full sun all the time. They are sad about this, but what's sadder is that they are wrong. It is true that most vegetables do best when planted in full sun, but there are plenty that do surprisingly well in shadier spots—although none should be planted in heavy shade.

WHAT HAPPENS IF YOU DON'T HAVE A GARDEN WITH FULL SUN

When the First Lady starts growing vegetables on the White House lawn, you know talk about the merits of growing at least some of your own food is on everyone's lips. It's not only fun to grow things you can eat, it helps us all understand a little better where food comes from and what it takes to grow it. So, even if you don't have the optimal amount of sun to work with, take heart! With the exception of sun lovers like tomatoes, cucumbers, peppers, squash, and eggplant, you can grow lots of delicious vegetables in shadier sites, as the table shows.

Leafy crops, like lettuce, mustard greens, arugula, endive, kale, cabbage, Brussels sprouts, Swiss chard, and many herbs tolerate shade best and can do with as little as two hours of direct sunlight. With just three to six hours of consistent, dappled sun you can grow beets, carrots, broccoli, cauliflower, onions, turnips, and radishes—though they will be more lush in sunnier locales. If you have a garden that offers both sun and shade, one of the advantages of knowing what can take less sun is the ability to increase your harvest by tucking shade-tolerant plants in along shaded borders that are often thought of as wasted space.

A BETTER WAY

Before you plant any vegetables, think first about what you and others in your household like to eat and can realistically eat in a season. Gardeners at every experience level are guilty of planting too many crops (it's easy to get carried away). Then you not only have to care for all the plants, you have to find ways to use them or give them away.

As far as care goes, watering consistently is always important to vegetable crops, but it's especially important to keep an eye on moisture levels in shady spots, which are often located beneath large trees or overhangs of the house or garage. Even if it rains, these spots may not get the water they need. So you will have to be the one to supply it.

—● the real dirt *There are plenty of vegetables that will grow in light shade.* Don't cheat yourself by not trying them. If you absolutely have no good spot in the yard, but have sun on a driveway or patio, you could put a few things you really want to grow in pots. Or you could put potted vegetables on some kind of rolling platform and roll them out into the sun. If you are home throughout the day, you can even roll them around to follow the sun as it moves.

Vegetables and shade tolerance

FULL SUN ONLY (6+ HOURS OF SUN)	WORTH A TRY IN LIGHT SHADE (4 TO 6 HOURS OF SUN)	CAPABLE OF HANDLING LIGHT TO MODERATE SHADE (2 TO 4 HOURS OF SUN)
Cucumber *(Cucumis sativus)*	Beans *(Phaseolus* spp.*)*	Arugula *(Eruca sativa)*
Eggplant *(Solanum melongena)*	Beets *(Beta* spp.*)*	Brussels sprouts *(Brassica oleracea* var. *gemmifera)*
Peppers *(Capsicum* spp.*)*	Broccoli *(Brassica oleracea* var. *botrytis)*	Cabbage *(Brassica oleracea)*
Squash *(Cucurbita* spp.*)*	Cabbage *(Brassica oleracea)*	Endive *(Cichorium endivia)*
Tomato *(Solanum lycopersicum)*	Carrot *(Daucus carota)*	Kale *(Brassica oleracea* var. *viridis)*
	Cauliflower *(Brassica oleracea* var. *botrytis)*	Leaf lettuce *(Lactuca sativa)*
	Coriander *(Coriandrum sativum)*	Mustard greens *(Brassica juncea)*
	Leek *(Allium porrum)*	Spinach *(Spinacea oleracea)*
	Onions *(Allium* spp.*)*	Swiss chard *(Beta vulgaris* ssp. *cicla)*
	Pea *(Pisum sativum)*	
	Radish *(Raphanus sativus)*	
	Rutabaga *(Brassica napus* var. *napobrassica)*	
	Turnip *(Brassica rapa* var. *rapa)*	

8
LAWN CARE

In This Chapter

Good Advice

∘ Spray water on dog spots to keep the grass from turning brown

∘ Do not sweep leaves, grass clippings, and other yard waste into the street

∘ Raise mower height from 2 to 3 inches for a healthier lawn

∘ Never cut more than one-third of the length of your grass at one time

∘ Spread grass seed only in spring and autumn

∘ Plant low-input grasses rather than traditional turf grasses

Advice That's Debatable

∘ Use sod rather than seed

∘ Use borax to control weeds

∘ Do not plant grass in the shade

∘ Allow animals and humans back onto treated lawns once the chemicals have dried

∘ Do not fertilize your yard because it contributes to environmental pollution

∘ Use organic fertilizers because they are more environmentally friendly

∘ Leave grass clippings on the lawn after mowing

∘ Water enough to keep grass green because a brown lawn is a dead lawn

Advice That's Just Wrong

∘ Use synthetic fertilizers and insecticides for lush, green grass

∘ Provide plenty of water to a lawn, especially in the heat of the summer

The most popular plant in American yards is grass. And, why not? You can walk on it, cut it to pieces, or even play football on it and it will grow back over and over again. Caring for grass is not that difficult, either. Fertilize once or twice a year, and water as necessary to keep it green. If you don't care whether your lawn is green, you can even skip the watering part and let it go dormant in the summer, when it will turn brown but not die—unless you live in the desert or have an unusually bad drought. Perhaps it is grass's popularity or its forgiving nature that makes it the subject of so much advice. As with other plants, though, some advice works and some, well, let's just say it's best to steer clear of it.

One of the fundamental problems with writing about how to care for lawns is that there are two distinctly different types of lawns: those composed of cool-season grasses, which grow well in regions where there is a long winter and winter temperatures regularly dip below freezing, and those composed of warm-season grasses, which do better in climates with short winters and temperatures dip below freezing only a few times during the year. There are some warm-season grasses that will grow well in cooler climates, but they tend to green up too late in the season for most people's preference, and many cool-season grasses do alright in warmer climates, just not as well as warm-season grasses. Many of the recommendations below revolve around cool-season rather than warm-season grasses, so keep that in mind as you peruse this chapter.

good advice Spray water on dog spots to keep the grass from turning brown

Dogs may just be grass's worst enemy. Having a dog pee on your lawn is one of the surest ways to turn that grass brown and crunchy. Female dogs have a reputation for being harder on grass than male dogs, because they squat when they pee so their urine pools on the ground in a concentrated puddle, whereas male dogs broadcast urine in more of a spray.

Dog spots occur not because of anything innately poisonous in the urine, but because there is a lot of nitrogen in the form of urea. While urea is also a popular fertilizer, it is so highly concentrated in urine it literally burns the grass. Have you ever noticed the ring of fast-growing, bright green grass that develops around dog spots? The nitrogen is more diluted away from the center of the urine pool, making it a fertilizer rather than a killer. In fact, if you dilute urine enough you can use it as fertilizer all around your garden (as we discuss in chapter 1).

WHAT HAPPENS IF YOU SPRAY WATER ON DOG SPOTS

If your dog (or neighbors' dogs) turns your grass into a sea of green punctuated by dead brown patches by urinating on it all the time, the only way to combat the problem is by diligently spraying the spot that gets peed on with water. This is not really feasible for every time a dog urinates. But try to reach for the hose when you see a dog urinate on the lawn to help keep brown spots somewhat under control. Studies by veterinarian A. W. Allard have shown that watering within eight hours of urination is best. Wait longer than that and it will be more difficult to beat the burn. Be aware, though, that even with spraying water you will have some spots, as well as areas that are greener than the rest of your lawn.

If you do not dilute the area where a dog has peed, and there is not any rain or an automatic sprinkler system to help your cause, you are probably going to have dog spots. Fortunately, there are plenty of companies selling dog spot patch kits, which include grass seed and some moisture-holding material, most of which work fairly well. You could also dig up the dead spot and seed the area. Still, diluting fresh dog pee with water seems to be the best option because it's less expensive and cures the problem more quickly.

HOW TO DO IT

Getting water to a dog spot is not difficult: just grab a hose or a bucket. The more water you apply, the more effectively you'll control the damage. But get that water on there as soon as possible and certainly within eight hours. In general, a gallon of water or so should do the job.

—➡ the real dirt *Dogs and pristine lawns do not mix.* But you knew that already. Still, you can keep your lawn from looking completely awful by spraying water on spots where dogs pee.

....

good advice Do not sweep leaves, grass clippings, and other yard waste into the street

Environmentally friendly types are constantly harping on the problems associated with sweeping things like leaves and grass clippings into the street. Instead, they say, we should put this material on our lawns, spread it on our gardens, or add it to our compost or yard debris recycling bins. As it turns out, they have got a good point. We know that grass clippings, leaves, and other plant debris releases nutrients as it decays. Nutrients from this decaying matter ultimately flow into lakes, streams, and other waterways, causing all kinds of problems for delicate aquatic ecosystems.

WHAT HAPPENS IF YOU SWEEP YARD DEBRIS INTO THE STREET

It's definitely easier to sweep or blow grass clipping, leaves, and other garden waste into the street, and years ago that is just what everyone did. But now that we understand more about what this decaying matter does to our environment, laws have been changed. Many communities require yard waste to be bagged and ready for pickup at the curbside or to be placed in yard debris bins for composting on a massive scale.

One great reason for leaving grass clippings on the lawn is that you can skip one fertilizer application each year. Grass clippings contain about 4 percent nitrogen, 2 percent potassium, and 1 percent phosphorus, so they provide a good portion of a lawn's nutritional needs. Leaves, too, are a great source of nutrients for soil and grass. Rather than raking fallen leaves, drive over them with a mower until they are shredded and spread them around as mulch.

When you rake grass clippings and garden debris into the street, you might as well be dumping it into a nearby lake. It's bad for the environment and bad for your lawn and gardens, too, because they could have benefitted from the addition of that organic matter. If you really have such an excess of grass clippings or leaves that you fear your grass may be killed in spots where you've left it, spread it around gardens, add it to a compost bin, offer it to neighbors—anything is better than putting it in the street.

HOW TO DO IT

Leave grass clippings on the lawn, rake them up and use them as mulch, or add them to the compost pile. You could also try using a mulching mower, which will make the clippings so fine you will notice them less. Leaves can also be mulched with a special mulching mower or just a regular old mower—just be careful not to pile the leaves too high or the mower will clog and shut off.

the real dirt *There is plenty of evidence that decaying grass clippings, leaves, and other garden debris left in the street will deliver nutrients to lakes, ponds, and streams that could harm them.* Besides, you'll spend less money on fertilizer and compost by spreading the organic material onto your lawn and gardens.

good advice Raise mower height from 2 to 3 inches for a healthier lawn

In the words of Nick Christians, a turf professor at Iowa State University, "Grass does not thrive on being mowed; it tolerates it." Truth be told, grass would prefer to be left unmolested if possible. But there are things we can do to soften the blow of mowing. Raising mower height is one of those things.

Hot, dry weather is really hard on both types of grasses, but particularly cool-season grasses. By leaving blades longer, grass is able to shade itself better and more efficiently produce food through photosynthesis. Higher mowing height also promotes a deeper root system that will help grass better hold up under drought stress. If you know what type of grass you have, you can even be fairly specific about the height that grass will best tolerate when mowed. In general, warm-season grasses usually do best when kept at a lower height than cool-season grasses, around 2 inches rather than 3 inches.

WHAT HAPPENS IF YOU DON'T RAISE THE MOWER HEIGHT

Mowing grass low does make a lawn nice and uniform, and it is tough on weeds. Unfortunately, it is just as tough on your lawn. If you choose to set the mower height lower and scalp your lawn, you are probably dooming yourself to more frequent applications of herbicide or more weeds, as well as a lawn that does not hold up well in times of drought.

HOW TO DO IT

Most hand mowers have height adjustment levers right next to their wheels, whereas on riding mowers the adjustment is usually on the mower itself. Simply move the lever to get the mower 3 inches off the lawn. You can do these adjustments on pavement if that makes it easier. The accompanying table lists the preferred mowing heights for several commonly planted grasses.

Preferred mower-height settings for grasses

Species	Preferred mower height
COOL-SEASON GRASSES	
Creeping bentgrass *(Agrostis stolonifera)*	low (1–2 inches)
Fine fescues *(Festuca spp.)*	medium (2–3 inches)
Kentucky bluegrass *(Poa pratensis)*	high (3 inches)
Perennial ryegrass *(Lolium perenne)*	medium (2–3 inches)
Tall fescue *(Festuca arundinacea)*	high (3 inches)
WARM-SEASON GRASSES	
Bermudagrass *(Cynodon dactylon)*	low (1–2 inches)
Buffalograss *(Bouteloua dactyloides)*	medium (2–3 inches)
Centipede grass *(Eremochloa spp.)*	medium (2–3 inches)
St. Augustine grass *(Stenotaphrum secundatum)*	high (3 inches)
Zoysiagrass *(Zoysia spp.)*	low (1–2 inches)

━● the real dirt *If you don't like the look of a longer lawn, try to get over it.* The reason is very straightforward: by adjusting mowing height upward, you will reduce plant stress and significantly improve the health of your lawn.

good advice Never cut more than one-third of the length of your grass at one time

As a rule, people who work with grasses recommend removing only about a third of a lawn's height (usually about 1 inch) when you mow. You will have to mow more frequently when you cut only this amount of leaf surface, but the payoff is worth the trouble. Cutting just one-third of grass height saves a lawn from stress that could damage the root system. This will also produce fewer clippings, so you can leave them on the lawn without having to worry about creating disease or thatch problems.

WHAT HAPPENS IF YOU CUT MORE THAN ONE-THIRD THE HEIGHT

A lawn is nice to have, but it can be a pain to care for. Proper mowing will keep a lawn healthy and looking good while reducing the amount of work you have to spend on it. If you regularly cut more than one-third off the top of your grass, especially in the heat of the summer, you run the risk of killing the grass or stressing it so much that it becomes diseased and needs to be watered often to survive.

HOW TO DO IT

It's best to set your mower height higher as opposed to lower. Generally, a height of about 2½ to 3 inches is recommended for cool-season grasses and 1½ to 2½ inches for warm-season grasses. Depending on the height of the grass, you may need to adjust the height of your mower so you are removing only about one-third of the leaf surface, about 1 inch, of grass at one time.

━● the real dirt *Cutting off only a third of the height of a lawn at one time may mean you have to mow more often, but you will be doing your grass and yourself a favor in the long run.*

....

good advice Spread grass seed only in spring and autumn

There is no question that the best times to plant grass are spring and autumn, but which season depends on the type of grass you are using. Warm-season grasses, such as zoysiagrass, Bermudagrass, and centipede grass, are best planted in the spring. Because they do most of their growing over the summer, seeding warm-season grasses in the spring helps them outcompete most weeds before going dormant in autumn. Cool-season grasses, such as fescues, ryegrasses, and Kentucky bluegrass, do best when they are sown in very late summer or autumn because they do most of their growing in slightly cooler weather and go dormant in the summer heat. When planted in autumn, cool-season grasses do not have to compete with as many summer weeds as they would if they were just getting started in the spring.

Another good reason for planting in spring and autumn is that rainfall is usually fairly regular at that time in most parts of the United States, which helps to get seedlings off to a strong start.

WHAT HAPPENS IF YOU DON'T SPREAD GRASS SEED IN SPRING AND AUTUMN

Grass seed can be planted almost any time during the year. Certainly, you should not hesitate to seed a bare patch of lawn in the middle of summer, as long as you care for it properly, of course. But nature is going to be more helpful to your cause if you seed during the spring or autumn, being mindful of the type of grass you are working with.

Seeding at other times, such as the winter or summer or during the wrong season for the type of grass, does not mean all of the grass seed will die. Much of it may do just fine. You will just have to be sure to water regularly as the seeds are germinating, and the grass will probably struggle more with weeds.

HOW TO DO IT

In general, cool-season grasses should be sown between mid-August and late September (maybe a bit earlier or later depending on where you live). If you must plant cool-season grasses in the spring, try to plant when the soil temperature is at least 40°F (5°C) or a little warmer. When that time comes depends on where you live, but will likely be between late February and early May. If you are planting warm-season grass, try to plant between March and late May, once the soil temperatures are 50° to 65°F

(10° to 18°C). Soil temperature is best measured with an inexpensive soil thermometer, which can be purchased at most garden centers.

━━● the real dirt *We have had to seed grass at a variety of times during the year.* With proper care all of these efforts have worked, but there's no doubt that it is best to plant warm-season grasses in spring and cool-season grasses in autumn.

....

good advice Plant low-input grasses rather than traditional turf grasses

Low-input grasses are sometimes called "greener" grasses, not because of their color but because maintaining them is easier on the environment. Although these grasses sometimes get a bad rap for their looks, it's really undeserved because there are plenty of low-input grasses that look good and can be walked or played on just like traditional grasses. Low-input grasses grow well with less water and fertilizer than more needy varieties, such as Kentucky bluegrass and perennial ryegrass. All warm-season grasses tend to need slightly lower inputs than cool-season grasses. Among the warm-season grasses, buffalograss *(Bouteloua dactyloides)*, zoysiagrass *(Zoysia)*, and Bermudagrass *(Cynodon dactylon)* fare best with lower inputs.

Some of the low-input cool-season grasses that make wonderful lawns are fescues, and there are four types we'd suggest planting: hard fescue *(Festuca brevipila)*, Chewing's fescue *(Festuca rubra* ssp. *fallax)*, sheep fescue *(Festuca ovina)*, and red fescue *(Festuca rubra)*. These fescues can get by with only about 1 pound of nitrogen per 1000 square feet applied in autumn compared to more conventional grasses that need about 1 pound of nitrogen per 1000 square feet applied twice or more per year. Fescues are not new. In fact, many seed mixes already include fescues, so if you don't fertilize heavily they will eventually dominate your lawn instead of the Kentucky bluegrass you planted. The same is true of warm-season grasses.

WHAT HAPPENS IF YOU PLANT LOW-INPUT GRASSES

In the long run, using lower-input grasses is better for the environment and easier on your pocketbook than growing Kentucky bluegrass, which looks nice and has a pretty name, but is a resource hog. If you don't go with a low-input alternative, you will have to do more watering and feeding to keep your lawn healthy and looking good. With a low-input fescue, you may miss the early- and late-season green-ups you get with a traditional grass. But, if you are not irrigating, you'll like the fact that in midsummer fescues don't go brown as quickly as other grasses do.

HOW TO DO IT

Making a change to a low-input fescue is simple. Just select one of the four fescues mentioned above and overseed your lawn, that is, apply the fescue seed over the top of your existing lawn at the rate recommend on the label. By cutting down on watering and fertilizing, the fescue will slowly take over the lawn. If you are putting in a new lawn, look for a seed mix with a high concentration of low-input fescues or shop specifically for a low-input mix.

—➤ the real dirt *It's difficult to understand why Kentucky bluegrass is still as popular as it is.* Sure it looks good, but low-input grasses, such as zoysiagrass in warmer climates and fescues in cooler climates, can look just as good and they take less effort.

advice that's debatable Use sod rather than seed

There's something attractive about buying grass that you can just unroll onto your lawn to get instant turf. Because sod looks so good when it is laid down, people expect it to do well. However, sod does have some inherent problems that you need to appreciate before you decide to use it. One of the most notable is that when you purchase sod it only has roots as deep as the sod is thick. So, until the grass develops a better root system, it will be extremely susceptible to drought. Additionally, if the sod you are using has a very different soil type than the ground over which it is being placed, its roots will not want to grow from nice topsoil (which the sod probably has) to, say, a heavy clay, leading to a shallow root system. Seeded lawns certainly are not immune to problems. When seed first starts to germinate, it is very susceptible to drought. Although seed is able to handle a broader range of soil conditions than sod, if it is spread over poor soil, it is still not going to fare very well. Both sod and seed tend to do best when the soil is lightly tilled before the sod or seed is applied. One of the greatest advantages of seed is that it is less expensive than sod.

WHAT HAPPENS IF YOU USE SOD

If you need grass to be green, lush, and beautiful quickly, there is nothing better than sod. Watered and cared for properly from the start, sod can look just as good or better than a seed-grown lawn. It will take longer to establish a lawn grown from seed, but the final result is usually the same: a nice lawn.

A BETTER WAY

Whether you are rolling out sod or planting seed into lightly tilled soil, you need to make sure that the type of grass you are using is right for your climate. Like seed, sod of warm-season grasses does best if it's rolled out in spring, and you should lay cool-season grass sod in autumn. If you must, you can put down sod in the summer, but you will need to water a lot to help it establish and grow during such extreme conditions. Both sod and seed will need to be watered a few times a week the first year they are planted. Fertilizing is usually best done after the grass is established, regardless of whether you use sod or seed.

━▶ the real dirt *If you are willing to pay extra to make your lawn green quickly and you realize that sod has shallow roots, sod can be a good choice.* If you have the patience to allow grass seed to germinate and establish a lawn, then seed is probably better.

....

advice that's debatable Use borax to control weeds

Of all of the weed control remedies that we have seen, borax is certainly one of the most interesting. It's also one of the most dangerous for your lawn. Borax really only works on a single common weed, creeping Charlie or ground ivy *(Glechoma hederacea).* Borax contains boron, an element needed by plants, though it can be toxic at high concentrations. Creeping Charlie is more sensitive to boron levels than most other plants, including grass. So you can apply boron to kill creeping Charlie without killing grass, but you have to be careful. If you add too much your grass will die, and you may end up with bare ground for two or three years until the excess boron leaches away.

WHAT HAPPENS IF YOU USE BORAX

If you want to try something new to control creeping Charlie, using borax might be a good experiment for you. We would not recommend trying it in the middle of your front lawn, because you could end up with a big bare spot. Make a borax spray by dissolving 10 ounces of borax in 4 ounces of warm water. Mix that with two and a half gallons of warm water and spray it over about 1000 square feet of creeping Charlie–infested area. Clearly you do not want pets or children around when you are doing this, and it would be best to wait a week to allow the mix to work its way into the ground before you allow them into the area.

A BETTER WAY

Because creeping Charlie grows best in moist, shady soil, you can discourage this weed by keeping your lawn dry and sunny. This can be achieved, for example, by aerating the soil, watering less often, and clipping overhead trees or bushes that shade your lawn. If the weed problem has not gotten too bad, pulling creeping Charlie out by hand is possible, but if you are dealing with a large area it can be extremely laborious. This plant is able to regrow from the smallest scraps, however, so when pulling it out be sure to get the entire root system and bag and remove all of the plant material so it doesn't infest your lawn again.

Aside from borax, there are other compounds that will control creeping Charlie. The most common way is to use the synthetic chemicals triclopyr, 2,4-D, mecoprop, or dicamba, all of which are best applied in autumn.

▬● the real dirt *Borax is not the amazing weed killer it is sometimes made out to be.* It kills creeping Charlie, but only when applied correctly. Use borax at too high a concentration and you'll end up with dead creeping Charlie and dead grass, too. We aren't saying you shouldn't try it, but be aware of the possible consequences before you do.

advice that's debatable Do not plant grass in the shade

What should you plant in the shade around your house? Common knowledge says that grass will not do well in the shade so you should select some sort of groundcover, such as Japanese spurge (*Pachysandra*). But grass does grow in the shade. Look around and you will see it in plenty of shady places. It just doesn't grow in really deep shade, like under a maple tree—though that often has more to do with dryness and lack of nutrients than shade, but we digress.

If you want to grow grass in the shade, you need to choose a grass that can tolerate it. Perennial ryegrass and Kentucky bluegrass, two of the more common lawn grasses, don't tend to do very well in anything darker than light shade, so strike them from your list. In cooler areas, fescues such as red fescue *(Festuca rubra)*, Chewing's fescue *(Festuca rubra* ssp. *fallax)*, hard fescue *(Festuca brevipila)*, and tall fescue *(Festuca arundinacea)* all perform reasonably well. In warmer areas, St. Augustine grass *(Stenotaphrum secundatum)*, *Zoysia*, centipede grass *(Eremochloa)*, and carpetgrass *(Axonopus)* are good choices.

WHAT HAPPENS IF YOU PLANT GRASS IN THE SHADE

While groundcovers are likely to look better than grass more months of the year, a lot of groundcovers do not hold up as well underfoot, so it's easy to see why people try to make grass work in less than optimal light. If you prefer grass to other groundcovers, you do have some options. You just need to be careful to choose shade-tolerant grasses suited to your climate.

A BETTER WAY

We generally prefer other groundcovers to grass because they require less water, fertilizer, and care. If you decide to forgo grass, you might try Japanese spurge *(Pachysandra)*, creeping phlox *(Phlox subulata)*, bugleweed *(Ajuga)*, sweet woodruff *(Galium odoratum)*, or even moss—and that is just the beginning of the list.

If you are dead set on grass, choose a shade-tolerant variety that is suited to your climate. Warm-season grasses should be planted in spring and cool-season grasses do best when planted in autumn. If grass is already planted in the area you are dealing with but it's not doing well, instead of tilling it up simply overseed the lawn with a shade mix. The shade-tolerant grasses will take over in time.

━● **the real dirt** *There's nothing wrong with growing grass in shade, and it can be done under the right conditions.* But we would suggest going with a different groundcover, if you can, because grass is not all that eco-friendly and it does not look great month after month, like many groundcovers do.

....

advice that's debatable Allow animals and humans back onto treated lawns once the chemicals have dried

Homeowners and the lawn care industry use a number of chemicals to control weeds. The most common of these include 2,4-D, mecoprop, dicamba, triclopyr, prodiamine, pendimethalin, and dithiopyr. Most of these chemicals are either soluble in water or have a surfactant (basically a soap) as part of their formulation. This means that water can actually carry these poisons across membranes, such as your skin. When grass is dry, the chemicals have a difficult time crossing into your body. This does not mean that the herbicides have ceased to exist. They're still there. They're just not as potentially dangerous as they once were.

WHAT HAPPENS IF YOU EXPOSE YOUR FAMILY TO LAWN CHEMICALS

Lots of people use chemicals or hire lawn care companies to use them to control weeds in their yards. Let's face it, spraying something is a lot easier than pulling weeds by hand. But is it safe to go back on the grass once the chemicals are dry? The companies who make and/or use the products say yes.

When it comes to 2,4-D, though, there may be cause for concern. Although the U.S. Food and Drug Administration deems the chemical safe, some studies have shown that it may pose a cancer risk. Because of this, countries including Norway, Denmark, and Sweden have banned the use of 2,4-D in lawns, and when we called several local lawn care companies asking about the chemical, almost all said they were phasing it out.

Despite what you may have heard or what you might assume, herbicides do not simply disappear when the grass dries. They are still there. Some studies have suggested that, even at the low levels at which they are found on a lawn, these chemicals may be toxic for us, our children, and even our pets.

A BETTER WAY

The best way to control weeds in your lawn is to fertilize and water your lawn properly, which will reduce the incidence of weeds. Where you do have weeds, hand pulling is always best. If you can't get all your weeds, what's a few dandelions between neighbors?

If you choose to use chemicals to control weeds in your lawn, always follow the directions on the label, especially those regarding how long to wait before entering an area after the application. If you are really cautious, you might even wait an extra day or two beyond what the label recommends.

━● the real dirt *Herbicides do not simply disappear when the grass dries, and some studies have shown that 2,4-D may pose a cancer risk.* After using chemicals on your lawn, carefully follow the instructions on the label regarding how long to wait before entering the area.

advice that's debatable Do not fertilize your yard because it contributes to environmental pollution

A long-standing point of debate is that by fertilizing your lawn you are contributing to nutrient-rich runoff that enters nearby lakes and streams. These nutrients turn bodies of water into dead zones, because they stimulate the growth of algae, which are then eaten by bacteria that uses up all of the oxygen in the water, thus suffocating fish and other aquatic creatures. This theory is flawed, though. There is strong evidence that when fertilizer is used properly and care is taken to prevent runoff, it helps grass develop a denser root system that catches most of the fertilizer.

We know it sounds unbelievable, but not using any fertilizer at all may actually be worse for the environment than using a modest amount. Fertilizing may include any of a host of nutrients. Nitrogen and phosphorus are the most important of these in terms of causing problems with aquatic ecosystems. In general, nitrogen applications should be made to lawns at least yearly, whereas phosphorus might not be needed at all. In fact, in some places it is illegal to add phosphorus to your lawn without a soil test showing a need for it.

WHAT HAPPENS IF YOU FERTILIZE YOUR YARD

Avoiding fertilizer out of concern for the environment sounds like a good idea. But lawns that do not get adequate nutrients develop weak root systems that require more water just to stay alive. Worse yet, undernourished grass has such low density that during a rainstorm water runs off of it faster than it would if it were in better shape. Even when lawns have an acceptable level of phosphorus in the soil, if they don't have the proper amount of other nutrients, such as nitrogen, runoff from a rainstorm could cause phosphorus-rich soil particles to wind up in waterways where they do not belong.

If you do fertilize your lawn but are concerned about the effects of runoff on nearby waterways, you could fertilize less often or use a smaller amount of fertilizer than is recommended. It may also be best—or even required by law in some places—to purchase only those fertilizers that do not include phosphorus.

A BETTER WAY

If you want to provide your lawn with some nutrients without using fertilizer, you can do this on smaller lawns by topdressing the grass with compost in spring and autumn (about ½ inch of compost will do). You could also rake fallen leaves onto the lawn in the autumn and use your mower to shred them. As they decompose, they will feed the soil.

If you just want something green for a lawn but don't really care what it is, let clover *(Trifolium)* take over or overseed your grass with clover. A common legume in lawns, clover gets nitrogen from the air and will provide surrounding grass with some nutrition without as much fear of runoff. This is not a good idea, though, if you enjoy being barefoot in your lawn and you're allergic to bees.

—● the real dirt *Unlike plants that grow less densely, grass does an excellent job of taking up the nutrients supplied to the soil around it.* Therefore, you can supply a lawn with somewhat less than the recommended amount of fertilizer and your grass isn't likely to die. Still, to get a lawn thick enough to prevent runoff of nutrients from the soil, your lawn does need to get nutrients from somewhere. If you are not going to use fertilizer, you need to have an alternate plan for feeding your grass.

advice that's debatable Use organic fertilizers because they are more environmentally friendly

This bit of advice is in the debatable section because it is not as plainly true as it sounds. Both organic and synthetic fertilizers can produce harmful runoff. The key is choosing the right type of fertilizer and using it appropriately for your situation. Organic fertilizers are said to be better for the environment because they release nutrients slowly. Quick-release fertilizers, which are often synthetic rather than organic, are very soluble in water, so they can be washed away more easily before they are fully taken up by plants.

But that makes this debate too simplistic. Like synthetic, quick-release fertilizer, organic fertilizer can wash away when it is applied too heavily or too frequently. When applied correctly, quick-release synthetic fertilizer is no more likely to produce nutrient-rich runoff than a slow-release organic fertilizer would.

The truth is, if you are applying any type of fertilizer to your lawn, you need to be prepared for some amount of it washing away. By assuming that organic fertilizers are inherently safer for the environment, homeowners may apply too much of them, creating more nutrient-rich runoff than they might have when using a synthetic fertilizer according to package directions.

WHAT HAPPENS IF YOU USE ORGANIC FERTILIZERS

If you are thinking of using organic rather than synthetic fertilizer, there are more convincing reasons to do so than the potential for producing less nutrient-rich runoff. Depending on the organic fertilizer you choose, it may be a better choice for the environment simply because it did not require the use of fossil fuels to produce, as many synthetic fertilizers do. Organic fertilizers can also be by-products, such as from the corn industry (corn gluten meal) or fishing industry (fish emulsion). Fertilizers containing organic material improve soil and, in turn, the long-term health of your lawn.

If you choose synthetic fertilizers over organic ones for your lawn because of lower cost, ease of use, or the instant gratification that comes from a lawn that greens up quickly, you can still do the environment a good turn by reducing the amount you use.

A BETTER WAY

Generally it's a good idea to apply about 2 pounds of nitrogen to every 1000 square feet of turf annually. Apply 1 pound in autumn, and 1 pound in spring. If you use synthetic fertilizers, slow-release types are the best choice.

If you want to use organic fertilizers in hopes of making a more environmentally friendly choice, you need to consider your options carefully. Try to stay away from fertilizers that include guano and rock phosphate, which are harvested or mined in ways that are not ecologically friendly. Instead use those that contain by-products, such as fish emulsions, seaweed extracts, or corn gluten meal. It takes more time and energy, but applying shredded leaves and compost to your lawn even semi-regularly will be helpful. Manure-based fertilizers are also a good choice.

—➧ **the real dirt** *Do not fall for the line that organic is always better.* Too much fertilizer used too often is a problem no matter what type it is.

advice that's debatable Leave grass clippings on the lawn after mowing

When mulching mowers became popular in the 1990s, there were some who felt that these were the worst things to ever happen to grass. Clippings, they said, would sit on the surface of the lawn and block sunlight, water, and nutrients. Worse yet, all those clippings would build up into a nasty thatch, that is, a buildup of plant debris, which can lead to pest and disease problems. They had a point, albeit a weak one. A big pile of grass clippings could certainly cause problems in a yard. But given how often most people mow, it's unlikely that you'd ever have such an abundance of them.

WHAT HAPPENS IF YOU LEAVE GRASS CLIPPINGS ON THE LAWN

Leaving clippings on your lawn can be a good thing for a lot of reasons. But too much of a good thing can cause problems. If the amount of grass left on top of a lawn is extensive, it is possible, though unlikely, that you could hurt some of the grass. If you do not leave clippings on your lawn for fear of depriving it of sustenance or creating a thatch problem, you will actually be depriving the grass of nutrients and plopping a valuable garden resource at curbside to be hauled away.

A BETTER WAY

Today's mulching mowers make it easy to mow and leave fine clippings behind to help feed grass slowly as it breaks down. After just a few days the clippings are not even visible. If you have areas of grass that are thicker and greener than others, spread those clippings around so you will not get too big of a pile lying around in one place.

━● the real dirt *Generally, it's a good idea to leave grass clippings on your lawn.* If there is a spot in your yard where grass grows faster than anywhere else, such as over a septic tank, you might need to rake some clippings out of that area.

....

advice that's debatable Water enough to keep grass green because a brown lawn is a dead lawn

Every summer, at least one person in every neighborhood lets the grass turn brown. People start grumbling about what a shame it is that they let their lawn die like that. But did the grass really die because they stopped watering it? Probably not. Grass is not like a tree. When a tree turns brown over the summer, it's more than likely dead. Grasses, particularly cool-season species, are built to turn brown.

Depending on where you live, over the summer grass may experience stretches of time when there is not much rain. When grass, particularly cool-season species, goes without water it does not die, it goes dormant. Different grasses stay dormant for different amounts of time, but they are all usually healthy while they are in this dormancy. Certainly, a prolonged drought will eventually kill grass roots, as well as the blades, but that does not happen very often. A month without a good rain or water of any kind might do it, though, depending on the grass.

WHAT HAPPENS IF YOU WATER THE LAWN OVER THE SUMMER

If you prefer to have a green lawn, you will need to water it over the summer, particularly in very hot weather. If you choose to allow your turf to become dormant, you probably will not have harmed its overall health. But even if you are willing to let your grass go dormant, you should still consider irrigating it once or twice during rainless periods during the real heat of summer just to be sure that it won't die.

A BETTER WAY

If you want to keep your grass green all summer, you will need to water whenever the grass starts to dry out. Whether from rainfall or an irrigation system, most lawns will need about 1 to 1½ inches per week in order to remain green.

━● the real dirt *If you want your grass to be green all year, you will probably need to irrigate.* Besides unsightliness, though, there's nothing wrong with letting grass go dormant over the summer. But if there's a dry spell lasting more than a month, you will need to get out the sprinkler if you want it to survive.

advice that's just wrong Use synthetic fertilizers and insecticides for lush, green grass

There's no doubt about it. The easiest way to get a lush, dark green lawn with no weeds is to douse your grass with fertilizers and weed killers. If you don't want to do it yourself, you can hire people to come over and do it for you. But this instant-gratification approach to lawn management is not without its problems, and it is absolutely not the only way to have healthy, good-looking grass.

In addition to other natural fertilizers we have mentioned, alfalfa and soybean meal (available at feed stores) are both good choices for giving lawns the nutrients they need. As far as weed control goes, corn gluten meal takes a while to work, and it's not quite as thorough as synthetic herbicides, but it does do the job. If you have a small lawn, hand weeding is not really that difficult. By taking care to improve the soil beneath your grass, you will have fewer weeds to pull in the future.

WHAT HAPPENS IF YOU USE SYNTHETIC FERTILIZERS AND INSECTICIDES

There are two major downsides to synthetic fertilizers. The first is that they offer only nutrients to a plant and don't provide any organic matter, like compost or renewable organic fertilizers, such as alfalfa meal or corn gluten meal, often do. The second is that synthetics fertilizers are not currently produced using sustainable methods. Their manufacturers often utilize natural gas or coal, which are nonrenewable resources. Synthetic pesticides have the same drawbacks that organic pesticides do. They can be toxic to the environment and to humans if they are used incorrectly.

A BETTER WAY

If you want a beautiful lawn, but are interested in caring for it in a more environmentally conscious way, you first need to have some patience. Building up good soil takes time. Using compost to increase the amount of organic material in your soil is an important step. Be sure to choose sustainable organic products suited for your yard and use them judiciously, because more is not better, even when something is organic.

━● the real dirt *Using the synthetic approach to lawn care is the easiest way to go, but it's not the direction that we hope most people choose to take because of the amount of non-renewable resources that it entails and because of the risks involved with using pesticides.*

advice that's just wrong Provide plenty of water to a lawn, especially in the heat of the summer

Overwatering is as dangerous to a plant's health as underwatering. Nevertheless, the myth that lawns can take as much water as you will give them persists. Just like other plants, lawns do have limits on how much watering they can endure before becoming predisposed to diseases or even dying.

Another thing that often gets overlooked is that trees commonly coexist with grass, so they can be added to the casualty list when a homeowner opts to ratchet up the lawn watering to a harmful degree. The area of the soil where grass roots reside drains quickly and the grass stays relatively healthy, at least for a while. But just a little further below, tree roots may be suffocating for lack of oxygen because the ground around them never dries out.

WHAT HAPPENS IF YOU PROVIDE TOO MUCH WATER
Whatever the circumstance, overwatering plants is a bad idea because their roots systems will suffocate.

A BETTER WAY

Overwatering a lawn is easier to detect than when overwatering some other plants because of the puddles and soft squishy spots that form in the lawn. These are a sign that you may be watering too much. If you want to keep your lawn healthy, stick to a schedule of no more than about 1 to 1½ inches of water every three days, including rainfall. Placing an empty tin can or a rain gauge in the lawn will help you measure much rain falls each week.

━● the real dirt *People think lawns need to be watered constantly.* In most cases, though, it's best to provide no more than about 1 to 1½ inches of water every three days or so to help build a stronger root system.

CONCLUSION

Polish astronomer Nicolaus Copernicus was not the first person to assert that the sun was likely the center of the solar system rather than the earth, as most scientists of the time believed. Like his predecessors, Copernicus' ideas were not well received, but he did not give up. His work spurred on additional scientific investigation, and though he died never knowing this, his theories helped lay the foundation upon which Galilei Galileo, Isaac Newton, and others built our modern view of astronomy.

The earth is flat. Nothing can travel faster than the speed of sound. Seemingly reasonable things are stated as fact every day, and when they get repeated they become stronger and seem more intractable. But most of these seeming truths are not resolute. Over the years, we gardeners have believed many things to be good practice that have proven otherwise: beat a tree to get it to flower, apply wound dressing to pruning cuts, use DDT to control insect pests, plant trees deeply to protect them from heavy winds—to name only a few.

If you take just one thing away from these pages, we hope it's the knowledge that you should always question what you hear. There are lots of reasons for the confusing, dubious, and just plain bad gardening advice out there. One of the biggest is that the people who are repeating it are ill informed. Maybe they are twisting the facts a bit. Maybe they don't know the facts. Perhaps they used to know their stuff but have not kept up with changing research.

We have done our best to help make clear what is true and untrue about the gardening advice we all hear most. And we have tried to shine a bit of light on some of the gray areas, where it's really a matter of personal preference or practicality rather than being right or wrong. But we are fully aware that time will change some of these things that we now believe to be true. So we will be watching, reading, researching, and questioning, and we hope you will be, too.

HELPFUL CONVERSIONS

Metric equivalents are approximate.

INCHES	CM
1/4	0.6
1/2	1.3
1	2.5
2	5.1
3	7.6
4	10
5	13
10	25
20	51

FEET	M
1	0.3
2	0.6
3	0.9
4	1.2
5	1.5
10	3
100	30
1000	305

WEIGHT	
1 ounce	30 g
1 pound	454 g
100 pounds	45 kg
500 pounds	225 kg

VOLUME	
1 tablespoon	15 ml
1 cup	240 ml
1 gallon	3.8 l

SELECTED BIBLIOGRAPHY

Appelhof, M., and M. Fenton. *Worms Eat My Garbage: How to Set Up and Maintain a Worm Composting System.* Kalamazoo, Mich.: Flower Press, 2006.

Bartholomew, M. *Square Foot Gardening: A New Way to Garden in Less Space with Less Work.* New York: Rodale, 2005.

Bird, R. *The Ultimate Practical Guide to Pruning and Training.* London: Hermes House, 2007.

Bradley, F. M., B. W. Ellis, and D. L. Martin, eds. *The Organic Gardener's Handbook of Natural Pest and Disease Control.* New York: Rodale, 2009.

Bryan, J. E. *Timber Press Pocket Guide to Bulbs.* Portland, Ore.: Timber Press, 2005.

Christians, N. *Fundamentals of Turfgrass Management.* 4th ed. Hoboken, N.J.: Wiley, 2011.

Damrosch, B. *The Garden Primer.* New York: Workman, 2008.

DiSabato-Aust, T. *The Well-Tended Perennial Garden.* Expanded ed. Portland, Ore.: Timber Press, 2006.

Druse, K. *Making More Plants: The Science, Art, and Joy of Propagation.* New York: Clarkson Potter, 2000.

Gillman, J. *The Truth About Garden Remedies.* Portland, Ore.: Timber Press, 2008.

Gillman, J. *The Truth About Organic Gardening.* Portland, Ore.: Timber Press, 2008.

Gros, M. *In Tune with the Moon 2010: The Complete Day-by-Day Moon Planner for Growing and Living in 2010.* Findhorn, Scotland: Findhorn Press, 2009.

Hartman, H. *Plant Propagation: Principles and Practices.* Upper Saddle River, N.J.: Prentice Hall, 1997.

Jeavons, J. *How to Grow More Vegetables (and Fruits, Nuts, Berries, Grains and Other Crops) Than You Ever Thought Possible on Less Land Than You Can Imagine.* Berkeley, Calif.: Ten Speed Press, 2002.

Jeavons, J., and C. Cox. *The Sustainable Vegetable Garden: A Backyard Guide to Healthy Soil and Higher Yields.* Berkeley, Calif.: Ten Speed Press, 1999.

Riotte, L. *Carrots Love Tomatoes and Roses Love Garlic: Secrets of Companion Planting for Successful Gardening.* Pownal, Vt.: Storey Publishing, 2004.

Snyder, L. C. *Native Plants for Northern Gardens.* St. Paul: University of Minnesota, 1991.

Stewart, A. *The Earth Moved: On the Remarkable Achievements of Earthworms.* Chapel Hill, N.C.: Algonquin Books, 2004.

Tucker, D. M. *Kitchen Gardening in America: A History.* Ames: Iowa State University Press, 1993.

Tudge, C. *The Tree: A Natural History of What Trees Are, How They Live, and Why They Matter.* New York: Crown Publishers, 2006.

INDEX

2,4-D, 200, 202

A

aerate, 202

alfalfa meal, 16, 41, 45, 212

algae, 205

alkaline copper quaternary compounds,
175–176

ammonium, 42–43

ammonium sulfate, 28

annuals, 104, 143

aphids, 182

arborvitae *(Thuja occidentalis)*, 150, 159

arugula *(Eruca sativa)*, 184

Asiatic lilies, 125

asparagus beetle, 182

Astilbe, 106, 116

B

bachelor buttons *(Centaurea cyanus)*, 106

bacteria, 43, 127

balled-and-burlapped trees, 140, 150

balloon flower *(Platycodon grandiflorus)*, 107

bark, 39

barrenwort *(Epimedium)*, 116

beans, 169

bearded iris *(Iris germanica)*, 106

bee balm *(Monarda didyma)*, 117

beets *(Beta)*, 184

beneficial insects, 80

Bermudagrass *(Cynodon dactylon)*,
195–196, 198

bindweed *(Convolvulus arvensis)*, 172

bisphenol A, 174

black-eyed Susan *(Rudbeckia fulgida)*, 125

black spot, 117

blanket flower *(Gaillardia aristata)*, 107

bleach, 127

bleeding heart *(Dicentra spectabilis)*, 116

blood meal, 83

blossom-end rot, 166–167

borax, 201

boron, 201

boxwood *(Buxus sempervirens)*, 159

Brassicaceae, 169

broccoli, 169, 180

Brussels sprouts *(Brassica oleracea* var.
gemmifera), 167

bugleweed *(Ajuga)*, 106, 203

bulbs, 104, 118

burlap, 147, 149

C

cabbage *(Brassica oleracea)*, 180

calcium, 166–167, 185

calcium nitrate, 42

capsaicin, 77

carbon, 16

cardinal flower *(Lobelia cardinalis)*, 107

carpenter ants, 98

carrots *(Daucus carota)*, 181

centipede grass *(Eremochloa)*, 195–196

Chewing's fescue *(Festuca rubra* ssp.
fallax), 199, 203

chives *(Allium schoenoprasum)*, 181

chromated copper arsenate, 175

Chrysanthemum, 107

clay, 16, 27, 51

clay worms, 61

Clematis, 125

clover *(Trifolium)*, 206

coconut coir, 30

columbine *(Aquilegia canadensis)*, 52, 106

companion planting, 70, 181

compost, 17–18, 89, 192, 206, 212

compost tea, 14, 50–51

conifers, 138

cool-season grass, 190, 195

copper sulfate, 74

coral bells *(Heuchera)*, 130

corn gluten meal, 16, 40, 45, 78, 208, 212

creeping bellflower *(Campanula rapunculoides)*, 171

creeping Charlie *(Glechoma hederacea)*, 201

creeping phlox *(Phlox subulata)*, 203

cucumber *(Cucumis sativus)*, 169, 177, 184

cucumber mosaic virus, 169

Cucurbitaceae, 169

D

daisies *(Chrysanthemum leucanthemum)*, 107

dandelions *(Taraxacum officinale)*, 172

daylily *(Hemerocallis)*, 52, 107

DDT, 85, 216

deadheading, 106

deer repellants, 73

Delphinium, 107

Dianthus, 107

dicamba, 202, 204

disease resistance, 117–118

dish soap, 77

dithiopyr, 204

dividing plants, 127, 137

dog spots, 191–192

dormant, 139, 156, 210–211

drainage, 30, 36, 51, 56, 62, 67, 184

drip irrigation, 63–65

E

earthworms, 14, 18–20

eggplant *(Solanum melongena)*, 166, 184

eggs, 73

eggshells, 166, 185

elms *(Ulmus)*, 142

endive *(Cichorium endivia)*, 186

Eucalyptus, 98

evergreen, 149

F

Fabaceae, 169

fencing, 83

ferns, 116

fertilization (spring and fall), 33

fertilizer, 145, 205

balanced, 48

organic, 207

synthetic, 204, 212

fescue *(Festuca)* 195

fireblight, 165

fish emulsion, 185, 208

fluorescent lights, 123

flush cut, 158

foam flower *(Tiarella cordifolia)*, 116

Forsythia, 159

full sun, 125, 185

Fusarium wilt, 117

G

garlic *(Allium sativum)*, 77, 181

geranium *(Geranium sanguineum)*, 107, 182

glyphosate, 81, 143

goat's beard *(Aruncus dioicus)*, 106

gourd, 169

grass, 190

clippings, 192, 209

gravel, 67

groundcovers, 202

grow lights, 123

guano, 45, 208

guard cells, 66

gypsum, 167, 185

H

hardening off, 109

hard fescue *(Festuca brevipila)*, 198, 203

hardiness zone, 108

heirloom vegetables, 164

high-intensity discharge lights, 123
Hosta, 106
hot sauce (Tabasco), 73
hybrid vegetables, 174
hydroponics, 56

I
incandescent lights, 123
insecticidal soap, 71
iron, 145
iron sulfate, 28

J
Japanese maple *(Acer palmatum),* 108
Japanese spurge *(Pachysandra),* 202–203
Joe pye weed *(Eupatorium purpureum),* 125

K
kale *(Brassica oleracea var. viridis),* 169, 186
Kentucky bluegrass *(Poa pratensis),* 195, 202

L
ladybeetles (ladybugs), 75, 80
lady's mantle *(Alchemilla vulgaris),* 106
lamb's ears *(Stachys byzantine),* 106
landscape fabric, 96–97
leafspot, 118
leaves, shredded, 94, 208
LEDs, 123
legumes, 43
lettuce *(Lactuca sativa),* 180
Liatris, 52, 125
lilac *(Syringa vulgaris)* 52, 126, 158
lime (dolomitic), 23, 26–27
lunar gardening, 179–180
lungwort *(Pulmonaria),* 106

M
magnesium, 23
manganese, 22, 26, 145
manure, 208

maples *(Acer),* 142, 148
mealybugs, 77
mecoprop, 202, 204
milorganite, 83
mites, 77, 80, 182
molasses, 50
moles, 83
moon cycles, 179
moss, 203
mowing, 194–194, 209
mulch, 17, 89, 94
mustard greens *(Brassica juncea),* 186
mycorrhizae, 14, 46–47

N
New England aster *(Aster novae-angliae),* 52
ninebark *(Physocarpus opulifolius),* 144, 160
nitrate, 42–43
nitrogen, 99, 131, 145, 208
    depletion, 100

O
onions *(Allium),* 181
organic material, 16–17
overhead watering, 62
overseed, 199, 203
overwatering, 213
oxygen, 56

P
peas *(Pisum sativum),* 180
peat, 30, 36
pendimethalin, 204
peonies *(Paeonia),* 106
peppers *(Capsicum),* 126, 180
perennial beds, 115
perennial ryegrass *(Lolium perenne),* 195, 202
perennials, 104, 115, 143
perlite, 30, 67

pesticide, 74, 212
pH, 22–28
phlox, 117
phosphorus, 131, 145, 193
pincushion flower *(Scabiosa columbaria),* 107
pine bark, composted, 30
pine needles, 49, 94
plantain *(Plantago major),* 172
planting, 111–112, 118–120
poppies *(Papaver),* 106
potassium, 131, 145, 193
    nitrate, 42
potato *(Solanum tuberosum),* 117, 169
potting medium, 67
powdery mildew, 62, 117
pre-emergent herbicide, 80, 172
prodiamine, 202
pruning, 158–160
    compared to topping, 153–154
    of fruit trees, 165–166
    and planting, 155–156
    of shrubs, 137–138
pumpkins *(Cucurbita),* 169
purple coneflower *(Echinacea purpurea),* 52
pyrethrum, 74

Q
Queen Anne's lace *(Ammi majus),* 106

R
rabbits, 83
radish *(Raphanus sativus),* 186
red fescue *(Festuca rubra),* 198, 203
red wigglers, 21
Rhododendron, 116
rock phosphate, 40, 208
rocks, 89
root-bound (pot-bound), 104, 111, 155
roots, 137, 140–143, 145, 151, 155, 213

rotating crops, 168
runoff, 205

S
salt, 145
salvia *(Salvia splendens),* 107
sand, 51
seaweed extracts, 41, 208
Sedum, 52, 129
seed, 123, 196, 200
shade, 185–187, 202
sheep fescue *(Festuca ovina),* 198
Siberian iris *(Iris sibirica),* 106, 125
slow-release fertilizer, 40, 89, 145, 185
snowberry *(Symphoricarpos albus),* 144
soaker hoses, 62
sod, 200
sodium, 22
soil moisture, testing, 60
soil test, 22, 100, 145, 168, 206
Solanaceae, 169
Solomon's seal *(Polygonatum multiflorum),* 116
soybean meal, 212
spacing plants, 121, 177
speedwell *(Veronica officinalis),* 107
spinach *(Spinacea oleracea),* 180
squash *(Cucurbita),* 177, 180
staking, 146
suckers, 165
sulfur, 27–28, 50, 72
sunscald, 147–148
sweet viola *(Viola odorata),* 106
sweet woodruff *(Galium odoratum),* 203

T
termites, 98
thatch, 209
thinning seedlings, 170

threadleaf coreopsis *(Coreopsis verticillata)*, 107

tickseed *(Coreopsis)*, 107, 125

tillage, 38–39

tobacco mosaic virus, 118

tomato *(Solanum lycopersicum)*, 117, 126, 166, 177, 184

tomato hornworm, 182

topping trees, 153

transplanting, 109–110, 129–130

treated lumber, 175

tree guards, 147

tree wraps, 147

triclopyr, 202

Trillium, 144

turf, 198, 208

turnips *(Brassica rapa),* 186

twine, 150

U

urea, 43, 191

urine, 73, 191

V

vermicompost, 20–21

vermiculite, 30–31, 67

Verticillium wilt, 117, 169

Viburnum, 160

vinegar, 127

Virginia bluebells *(Mertensia virginica),* 144

voles, 83

W

warm-season grass, 188, 197

watering, morning, 59

water sprouts, 138, 153

weevils, 182

white and blue clips *(Campanula carpatica* 'White Clips' and 'Blue Clips'), 107

wilt, 66

winterberry *(Ilex verticillata),* 144

wood chips, 17, 89

wound dressing, 154

Y

yarrow *(Achillea millefolium),* 106–107

Z

zinc, 145

zoysiagrass *(Zoysia),* 195–198, 203

ABOUT THE AUTHORS

photo by Chad Giblin

Jeff Gillman is an associate professor in the Department of Horticultural Science at the University of Minnesota. His previous books include *The Truth About Garden Remedies, The Truth About Organic Gardening,* and, most recently with Eric Heberlig, *How the Government Got in Your Backyard.* His scholarly publications cover a broad range of topics, such as spider mite control, soil amendments, and treating plant diseases using organic means.

photo by Mike Hoium

Meleah Maynard is a journalist, editor, and master gardener. She writes regularly for regional and national magazines, and her garden and horticulture stories have appeared in magazines such as *The History Channel Magazine; Gardening How-To; Garden, Deck, and Landscape; Midwest Home;* and *Northern Gardener.* Meleah is also a garden columnist for the Minneapolis *Southwest Journal* and *Northern Gardener* magazine.